NORTH CAROLINA
STATE BOARD OF COMMUNITY COLLEGES
LIBRARIES
ASHEVILLE-BUNCOMBE TECHNICAL COMMUNITY COLLEGE

Discarded
Date

JUN 23
2025

Search for a Supertheory

From Atoms to Superstrings

Search for a Supertheory
From Atoms to Superstrings

Barry Parker

Plenum Press • New York and London

Library of Congress Cataloging in Publication Data

Parker, Barry R.
　　Search for a supertheory.

　　Includes bibliographical references and index.
　　1. Unified field theories. 2. Particles (Nuclear physics) I. Title.
QC173.7.P364　1987　　　　　　　　530.1′42　　　　　　　　87-13340
ISBN 0-306-42702-8

First Printing—September 1987
Second Printing—March 1989

© 1987 Barry Parker
Plenum Press is a division of
Plenum Publishing Corporation
233 Spring Street, New York, N.Y. 10013

All rights reserved

No part of this book may be reproduced, stored in a retrieval system, or transmitted in any form or by any means, electronic, mechanical, photocopying, microfilming, recording, or otherwise, without written permission from the Publisher

Printed in the United States of America

Preface

What is the universe made of? What forces hold it together? And how are they related? These are questions that have puzzled scientists for decades. Today, scientists believe they are close to an answer. It is not yet a complete answer, but tremendous strides have been made. *Search for a Supertheory* is the story of what these advances are and how they were made. But it is also the story of the scientists involved in the search, their frustrations, hardships, hopes, and joy when great discoveries are made. It traces the advances in particle physics from the discovery of the atom and its components through to today's most exciting theory—Superstring Theory. And although it has the same theme as my earlier book *Einstein's Dream*, namely, the search for unity, the emphasis is quite different. In that book I emphasized the macrocosm; in this book I emphasize particles and fields. In the early chapters of the book you may, in fact, be overwhelmed by the large number of seemingly unrelated particles that have been discovered. This is, of course, exactly the way scientists felt at the time. But as you continue you will see how everything eventually came together and began to make sense. And you will also, I hope, share some of the excitement that physicists are now experiencing as the last pieces of this great scientific adventure are being put in place.

    There is no mathematics in the book but it is impossible to talk about science without using scientific terms, and it is likely that you will not be familiar with some of them. I have attempted to define each of these terms as it appears, but for the benefit of those who are new to science I have added a glossary. Very large and very small numbers are also needed occasionally and I have used scientific notation to designate them. The notation 10^{30}, for example, represents the large number one with 30 zeros after it. Similarly, 10^{-30} represents the small number one divided by 10^{30}.

Also used extensively is the energy unit, the electron volt. It is the energy an electron acquires in moving through a potential difference of one volt—roughly the voltage of a flashlight battery. Most of the accelerators discussed accelerate particles to millions (MeV) or billions of electron volts (GeV).

The sketches of the physicists were done by Lori Scoffield,* and the line drawings by Sandra Carnahan. I would like to thank both of them for an excellent job. I would also like to thank Murray Gell-Mann for several helpful suggestions and Julius Wess for the interview. And finally I would like to thank Linda Greenspan Regan, Victoria Cherney, and the staff of Plenum Publishing for their assistance in bringing the text into its final form.

Barry Parker

---

*The sketches of the physicists were adapted from Weber, White, and Manning, 1956, *College Physics*, with permission from McGraw-Hill.

Contents

CHAPTER 1 Introduction   1

CHAPTER 2 Probing the Atom   15

    Bohr and the Nuclear Atom   24
    The Discovery of Quantum Mechanics   29
    Discovery of the Neutron   37
    Yukawa and the New Particles   42

CHAPTER 3 Particle Accelerators   49

    Lawrence and the Accelerator   49
    Detectors   57

CHAPTER 4 Organizing the Particle Zoo   59

    The Eightfold Way   69

CHAPTER 5 Overcoming Infinity   77

CONTENTS

CHAPTER 6 Building a Universe   97

"Quark, Quark"   97
A New Beginning (the New Physics)   107

CHAPTER 7 Gauging the Universe   117

The Weak Interactions   125
Parity   130
Symmetry Breaking   140
The Weinberg–Salam Theory   143

CHAPTER 8 Adding Color   149

Freedom and Slavery   152
QCD   158
Imprisonment   160

CHAPTER 9 Adding Charm   165

Charm   169
The SPEAR Experiment   175
Properties of the J/ψ 181
The Charmonium Spectrum   182
Naked Charm   184
The Tau Lepton   186
Upsilon   187
Summary   189

CHAPTER 10 Search for the W   191

Jets   191
Search for the W   196

Success at Last   *201*
The Higgs Particle   *203*

CHAPTER 11   Unifying   *205*

Proton Decay   *214*

CHAPTER 12   Looking Deeper   *217*

Technicolor Theory   *218*
Preons   *219*
Rishons   *221*
Problems   *224*

CHAPTER 13   Supergravity   *227*

CHAPTER 14   Adding More Dimensions   *237*

Kaluza–Klein Theory   *237*
Modern Theories   *241*
Higher Dimensions   *242*
Supergravity   *244*

CHAPTER 15   Superstrings: Tying It All Together   *247*

CHAPTER 16   Cosmic Strings and Inflation   *261*

Grand Unified Theory to the Rescue   *264*

Inflation   *266*
Cosmic Strings   *270*
The Ultimate Question   *272*

CHAPTER 17   Epilogue   *275*

Glossary   *277*

Further Reading   *285*

Index   *287*

CHAPTER 1

Introduction

In the last few years scientists have been sifting through an ocean of scientific facts, organizing them, trying to make sense out of them, trying to extract from them an ultimate understanding of nature. And their efforts are finally beginning to pay off; they have brought us to the verge of one of the greatest breakthroughs science has ever seen. We are, even now, getting the first glimpse of a theory that will show us in detail how the universe came into being, what it is made of, and how it is put together. Once achieved, this theory will unify all of nature, from gigantic clusters of galaxies down to tiny elementary particles. It will unlock secrets of the universe we never dreamed possible. In short it will give us the master plan—the blueprint—of the universe.

Excitement is running high in the world of high-energy physics as we close in on this goal. Particle physicists are working around the clock, stretching their imaginations to the limit in an effort to make things fit, setting up ever more complex experiments in hopes of finding the last vital pieces of the puzzle.

At one time it was thought that the atom was the ultimate building block of matter. But as small as atoms are, there are particles that are a million trillion times smaller. If you could construct a microscope powerful enough, you would see that the atom is mostly empty space, made up of a tiny nucleus consisting of protons and neutrons, surrounded by a cloud of electrons. The protons and neutrons are small, but the electrons are far smaller. As strange as it seems, they may take up no space whatsoever.

The secrets that physicists are after involve not the atom, but its building blocks—the electron, proton, and neutron. At one time it was thought that all three of these particles were elementary in the sense that they are "ultimate" building blocks, but we now know this is not the case. Inside the protons and neutrons are even smaller particles called quarks. Evidence for these quarks was found at Stanford University's large accelerator near Palo Alto, California, in 1969. Realizing that protons were much larger than electrons, and might have hidden structure, scientists decided to bombard them with electrons. In a sense it was a rerun of a famous experiment performed in 1911 by Ernest Rutherford in which alpha particles (helium nuclei) were projected at gold atoms. To Rutherford's surprise some of the alpha particles were deflected through large angles. This was inconsistent with the accepted atomic model of the time, and Rutherford soon realized a new model was needed in which the electrons orbited a tiny, but heavy nucleus. In the Stanford experiment similar unexpected deflections occurred; this meant there had to be small pointlike particles inside the proton.

But Murray Gell-Mann of Caltech, and independently George Zweig at CERN, had already suggested that protons were composed of more elementary particles. In fact it was Gell-Mann who gave them the whimsical name "quarks," from a line in *Finnegans Wake*, "Three quarks for Muster Mark!" Consistent with the quote there were three kinds of quarks in each proton and neutron. Coincidentally, there were also three kinds of quarks in Gell-Mann's scheme; he called them up (u), down (d), and strange (s). But soon, as other discoveries were made, it was found that three was not enough. First, with the unexpected discovery of a particle we now call J/psi came a fourth quark, which was named charm. Then came two others, called bottom and top, for a total of six.

How many quarks are there? We do not know for sure, but it appears that this may be the end of the line. Inconsistencies develop in cosmology—the study of the structure of the universe—if there are many more. Does this mean, then, that the universe is built entirely of quarks? No; if we look back at the electron we see that it isn't composed of quarks, yet it is elementary. In fact, just as there is a quark family, so too there is an electron family, or as it is usually called, a lepton family. And just as there are six members of the quark family, so too there are six members of the lepton family. Besides the electron there is a particle that is similar to it except that it is heavier; we call it the muon. The third member is also like the electron, but it is heavier yet; it is called the tau. Corresponding to each of these particles is an elusive particle that may have no mass; it is called the neutrino.

For a while the quark model seemed to solve the problem of the ultimate structure of matter, but eventually difficulties developed and a new concept called color had to be introduced to overcome them. Color, as we refer to it here, though, has nothing to do with the usual meaning of the word. Certainly quarks wouldn't be red or blue if we could see them. It is a property of quarks, similar to electric charge, that enables them to join together to form particles such as the proton.

In short, then, it appears as if our world is made up of twelve different types of particles, six of which can be colored. But, strangely, even this isn't the end of the story. Back in 1928 the English physicist Paul Dirac predicted that to every type of particle in the universe there was an antiparticle, and when an antiparticle and a particle met, they annihilated one another in a burst of energy. Shortly thereafter, the positron, a particle similar to the electron except that it had a positive charge (the electron has a negative charge) was discovered, verifying his prediction. This means that in addition to the families of quarks and leptons there are also families of antiquarks and antileptons.

But these particles have to be held together if we are to have nuclei, atoms, and large objects such as stars. And indeed the forces that hold them together have been known for some time. The best known is gravity, the force that holds you to the earth.

But there's also a force that holds atoms together, called the electromagnetic force, and within atoms we have neutrons and protons held tightly together in the nucleus by a force called the strong nuclear force. The last of the four forces, one we'll talk about in more detail later, is called the weak nuclear force.

An important breakthrough in our understanding of these forces came in the late 1960s. It was found that the electromagnetic and weak nuclear force could be described by the same theory, and furthermore, that they could be mixed together. Although it was not a true unification, it showed us that fields could be joined, and it soon led to an attempt to understand the other fields in terms of unification. Unification, as we will see throughout this book, is one of the great quests of physics. The participants in the quest are theorists and experimentalists. The theorists devise theories and make predictions; the experimentalists build equipment and carry out experiments to see if the predictions are correct. If not, the theory is discarded and a new one is found to replace it.

Dramatic developments have capped our search for unification. Physicists have found that the macrocosm—the universe—and the world of elementary particles are intimately related. The elementary particles and forces of nature as we understand them today, were "fused" during the first fraction of a second after the big bang explosion that created the universe.

The big bang was like a giant "atom smasher," or particle accelerator, as it is referred to by scientists. But the energies and speeds generated by it were far greater than anything we can create here on earth. It is because of this link that physicists are trying to build larger and larger accelerators; they hope to probe ever closer to the first moment of creation. Much of our knowledge of elementary particles has, in fact, been made possible by the use of accelerators.

Although you may not realize it, you have a small accelerator in your home—your TV set. In a TV set a beam of electrons is accelerated toward the screen using high voltages. Magnets direct the beam, sweeping it back and forth across the screen, creating a picture. High voltages and magnets are also important in the giant accelerators scientists use. The first accelerators were built in the early 1930s by John Cockcroft and Ernest Walton in England, and by Ernest Lawrence in the United States. While Cockcroft and Walton's accelerator worked on the same principle your TV does, Lawrence's instrument was quite different. His first instrument looked like a pillbox, about 4 inches across. Charged particles spiraling around inside this box were given a boost each time they passed a certain point.

One of the first goals was to produce a machine that would give charged particles an energy of one million electron volts (an electron volt is roughly the energy gained by an electron as it flows from the positive to negative terminal of a flashlight battery). Lawrence and Livingston were the first to achieve the goal, beating out Cockcroft and Walton, but in their race for higher energy they lost an even more important race. Cockcroft and Walton had considerably less energy available but they used their instrument where it counted—for experimentation. And lo and behold in one of their first experiments they observed nuclear disintegration. They lost the race for a million electron volts, but they won where it really counted—in the physics.

But in the long run it was the cyclotron, the accelerator created by Lawrence, that was to become the important probe of nature. Lawrence and Livingston continued building larger and larger accelerators, and soon others were also building them. About twenty were completed before World War II.

The war halted cyclotron production but it brought an important new concept: that of a "scientific team." Teams of scientists were set up at Chicago, Los Alamos, and Oak Ridge, and after the war confidence was so high, with the success of the atomic bomb, that it was believed such teams could do almost anything. National laboratories were soon set up. Lawrence Laboratory in California easily had the lead in cyclotron development, but an eastern equivalent—Brookhaven National Laboratory—was soon built.

And with these new labs and their accelerators came an influx of discoveries. At first everyone was excited as new particle after new particle was discovered. But eventually there were so many different types of particles physicists were overwhelmed. Willis Lamb, in his Nobel address, said, "The finder of a new particle used to be awarded a Nobel prize, but such a discovery now ought to be punished by a $10,000 fine." Most of the new particles were considerably heavier than the electron and proton, and all were extremely short-lived. We now know that these particles are not truly elementary particles, but are composed of quarks.

In the race to build large accelerators Europe was determined not to be left behind. In 1952 several European nations banded together and created CERN near Geneva, on the Switzerland–France border. Although they were years behind the United States when they began, they soon caught up. By 1962 Europe matched the United States in scientific manpower but by the late 1970s it had nearly 3000 particle physicists—twice the number the United States had. And its persistence eventually paid off. The quest for the elusive particles called W and Z that occur in the radioactive decay of atoms was a goal both groups were striving for in the early 1980s. They were found at CERN in 1983.

The Russians were also soon in the race. A facility similar to CERN for socialist countries was set up shortly after the war at Dubna, near Moscow. By 1954 they had completed a large synchrocyclotron (modified cyclotron). Americans were amazed at the progress the Russians had made—the synchrocyclotron was nearly twice the size of the Berkeley machine.

Following Brookhaven other facilities soon sprang up around the United States. A huge, two-mile accelerator, designed to accelerate electrons in a straight line, was built at Stanford University (called SLAC), and later a similar accelerator

Measuring particle tracks at Fermilab. (Courtesy Fermilab.)

Aerial view of the Alternating Gradients Synchrotron (AGS) at Brookhaven. (Courtesy Brookhaven National Laboratory.)

was built at Los Alamos. Then in 1967 America capped off its efforts with Fermilab at Batavia, Illinois.

The innocence of small-time physics was now long-gone. Where scientific experiments could once be done with small machines that stood on a desk, accelerators now snaked through underground tunnels several miles long. And where there were once three or four people working on a project there were now hundreds. The list of names on some scientific papers was almost as long as the paper itself.

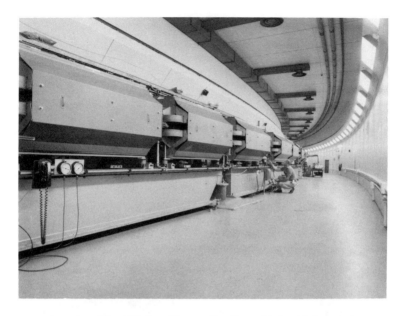

A portion of the AGS ring. (Courtesy Brookhaven National Laboratory.)

CERN's particle accelerator lies 15 stories underground. Thousands of magnets guide the stream of protons along its 4-mile length. When experiments are in progress everyone is cleared from the area around the accelerator because of the lethal radiation generated. The experiment is conducted from a large control room. The beam, traveling at a speed near that of light, is directed into a target. The devastating collision generates dozens of new particles. It's like having two billiard balls collide at high speed with eight or ten new ones coming out of the collision. But these new particles are all short-lived, decaying almost immediately to lighter particles.

Similar experiments are being conducted around the clock at Fermilab where a new accelerator is able to accelerate particles to a trillion electron volts. This new accelerator, called the Tevatron, was the first to use superconducting magnets—

The SUPERHILAC accelerator at Lawrence Berkeley Labs. (Courtesy Lawrence Berkeley Labs.)

An aerial view of Fermilab. The large circle is the main accelerator. Wilson Hall can be seen along the circle. (Courtesy Fermilab.)

special magnets that require almost no electricity to keep them running. The central building at Fermilab is an impressive 16-story structure near the main entrance to the lab. Two thousand full-time employees work here, and there are also at least 2000 visiting scientists here at any time.

Particles are first accelerated in a small accelerator to an energy of about 2 million electron volts. From here they are injected into a larger accelerator that boosts them to 8 billion electron volts, then they are passed to the upper ring of the main accelerator where 1000 water-cooled magnets direct them

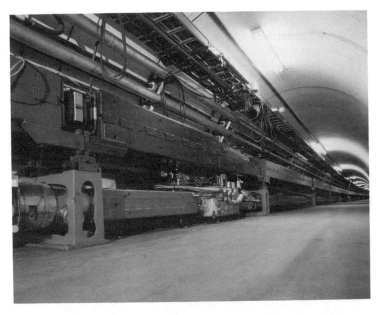

The tunnel of the main accelerator at Fermilab. (Courtesy Fermilab.)

in a circle about 4 miles in length. Finally, they are transferred to a ring of 1000 superconducting magnets directly below this.

While the United States, with the Tevatron, presently has the lead in energy, other countries are on their heels. The USSR hopes to bring an accelerator into operation in the late 1980s that will generate 3 trillion electron volts. A similar machine is presently being built at CERN. Other countries are also getting into the act. The DESY facility in Hamburg (West Germany) and KEK in Japan are both capable of making major breakthroughs in high-energy particle physics.

The success of superconducting magnets at Fermilab has spurred the United States on to the planning stages of an even greater giant—the superconducting supercollider (SSC). It will take the form of a loop about 60 miles in length. Inside the loop two pipes will contain protons moving at near the speed of light; in one pipe they would move clockwise, in the other anticlockwise. This model will be capable of accelerating protons to 20 trillion electron volts.

Electromagnetic fields will be used to give the protons a slight boost each time they orbit the huge loop, directed by superconducting magnets. Several thousand of these specially designed magnets will be needed to curve them into the required path.

It will take at least 4 billion dollars to build the SSC, and it will be 10 years before it will be completed. But it is expected to be 20 to 40 times as powerful as anything we now have.

Why do we need the SSC? Earlier I talked about an important new breakthrough in science—a unified theory of the universe. We are, indeed, getting close to attaining this theory. Large accelerators such as the SSC will help us achieve our goal. We saw earlier that the universe is made up of quarks and leptons, and these particles are held together by four basic forces. One of the major objects of physics is to show that these four forces are related to one another—under different conditions they are a single universal superforce. As I mentioned earlier, part of this goal has already been achieved: the electromagnetic and weak forces have been joined to form the electroweak field. Furthermore we have a tentative theory in which the strong nuclear field is joined to this electroweak theory. But we're not certain it is a correct theory. To find out we must perform experiments, and this is where large accelerators are important. Only through them will we achieve the understanding we are striving for.

Einstein searched for a unified field theory on a modest scale for the last years of his life, but never found one. We are now, however, getting close to this goal. Several schemes are currently at the forefront of our search. One is called grand unification theory (GUT), another is called supersymmetry. Both are based on symmetry, the property an object (or process) has if it does not change when certain operations are performed on it. A ball, for example, is symmetric with respect to rotation, in that it looks the same after we rotate it. One of the most important breakthroughs in recent years has been the realization that symmetry is extremely important in nature. Particles have been linked to one another in symmetric families. But there are also deeper and sometimes "hidden" symmetries at work. An important symmetry known as gauge symmetry has led the way in recent years. There may also be a broken symmetry called supersymmetry. It predicts a large number of particles that we have not yet seen. But perhaps with the SSC we will see them.

Even stranger ideas have recently been put forward. According to one, we may be living in a world of ten dimensions, six of which are unobservable. Furthermore, particles may not be the pointlike objects that we usually think of them as; they may be tiny vibrating "strings." Which of these ideas is correct we do not yet know, but the more we probe into the complicated mathematical structure of the theories, the closer we get to nature's deepest secrets, and the closer we get to our ultimate objective: a unified theory of the universe—a "supertheory."

In order to understand what this supertheory is all about, though, we must begin at the beginning and work our way forward one step at a time. In this chapter we had a brief overview of the particles and fields that make up our universe. In the chapters that follow we will look at each of these particles and fields in detail. And we will see how our understanding of them gradually grew, how the fundamental particles were first placed in groups, or families, with similar properties, how we discovered quarks, how the electromagnetic field was joined with the weak nuclear field, and how the strong nuclear field was joined with them. And finally we will look at recent attempts to unify all of nature.

CHAPTER 2

Probing the Atom

Although early Greek philosophers argued that the world was made up of small, indivisible particles, it was the English chemist John Dalton (1766–1844) who gave us our first modern atomic theory. Despite his great contribution to science, Dalton was, oddly enough, not a highly skilled scientist; he was actually quite inept in the laboratory. But it is, after all, ideas that count—not how you arrive at them. He never married, perhaps because he never had time; his daily routine was so rigid that there was very little time left over. He did lawn bowl occasionally, but in general was a dull, colorless individual with a single passion: weather. Each day he would dutifully record the temperature, rainfall, cloud cover, and so on. But eventually, as a result of this passion, he began to wonder: What is air made of? Why do clouds form in it? And this sense of wonder led him, in turn, to the lab where he began experimenting, not just with air, but with many different gases—hydrogen, oxygen, carbon dioxide. He weighed them; he mixed them and weighed them again, then he measured their pressure. Soon an idea began to take shape: gases were made up of small, indivisible units—atoms. In 1808 he published his theory in a two-volume book titled *A New System of Chemical Philosophy*.

Dalton's idea eventually caused a revolution in science, and it brought him considerable glory, even in his own lifetime. Among these honors was a doctorate awarded by Oxford in 1832. When he found out he would have to wear scarlet robes to

CHAPTER 2

John Dalton.

receive the degree, though, he was shocked. He was a Quaker, and scarlet was a forbidden color. Fortunately, there was an easy solution. He was color-blind; the robes appeared gray to him, and if he couldn't see their color he wasn't disobeying the church. He wore them and received the degree.

By the time of his death in 1844 Dalton was famous throughout Europe. During his lifetime he shunned honors, but he could do little about them after he was dead, and as a result his funeral was made into an elaborate last tribute to him.

At this stage the atom was still a tiny, structureless, indivisible piece of solid matter. But was it really structureless? Perhaps it was made up of more elementary particles. Almost 50 years passed before scientists found that this was indeed the case.

Faraday had set the stage. Because of his famous experiments on electricity and magnetism much was known about electrical phenomena, yet, strangely, scientists were still not

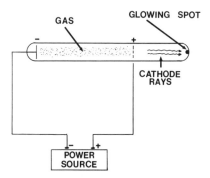

A cathode ray tube similar to the type used by Thomson.

sure what an electrical current was. They could see such a current in one of the major apparatuses of the day—the cathode tube, an evacuated tube with two terminals implanted in it that could be attached to the terminals of a battery. When the terminals were attached a strange glow appeared. And it was soon noticed that the glow changed color when a different gas was introduced into the tube. Furthermore, it was noticed that if you cut a small hole in the positive terminal (the anode) the beam continued on through it and struck the glass at the far end of the tube.

What was this beam? Some thought it was a strange new type of light; others were convinced it was a beam of particles. But there was a problem: while magnetic fields easily deflected the beam, electric fields seemed to have little effect on it. To J. J. Thomson of Cavendish Laboratory this was the key. He set up an apparatus and carefully measured the deflection in both electric and magnetic fields. He soon convinced himself that the beam consisted of particles, and that the small deflection in electric fields was due to the high speed of the particles. Then, from the direction of their deflection, he was able to determine that they were negatively charged. Finally, combining electric and magnetic fields in the same experiment he was able to ob-

CHAPTER 2

J. J. Thomson.

tain their speed and the ratio of their mass to their charge. The result was amazing: their speed was a tenth that of light, and their mass to charge ratio was 2000 times as great as that of hydrogen.

In 1897 Thomson announced his discovery. He named the particles electrons, and explained that electric currents were made up of millions of these tiny particles. But he did not stop here. He realized that the electron had to be part of the atom. There were, after all, atoms of gas in the tube, and the electrons had to come from them. His next step was to build a "model" of the atom, and he soon had what he referred to as a "plum pudding" model. It consisted of negatively charged electrons in a positively charged background; the background was his "pudding." Light was emitted by the atom when the electrons oscillated. It was, in many respects, a reasonable model for the time, but there were many things it couldn't explain. In fact, when it really came down to it there were few things it could explain.

One of the major things it couldn't explain were spectral

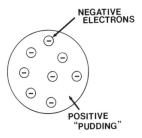

Thomson's plum pudding model of the atom.

lines. These were lines that occurred when the light from various hot gases (e.g., hydrogen, oxygen) was passed through a slit, then through a prism. They were colored lines characteristic of the gas, that always appeared in the same position and therefore could be used to identify the gas.

Problems such as this plagued Thomson's model but what else was there? Almost nothing at that time. There was a model that few talked about—one in which the electrons moved in orbits, much as the planets orbit the sun. No sooner had this model been suggested than someone pointed out a serious flaw. It was well known that charged particles emitted radiation when they accelerated (and in the process lost energy). And the electron was a charged particle; furthermore, when in orbit it was also accelerating. This meant that an electron in an atom would spiral rapidly into the center of an atom as it radiated away its energy. In fact, when the calculations were performed, it was found that it would take less than a millionth of a second for this to happen. Needless to say, interest in planetary models soon waned.

But planetary models didn't die altogether, and a discovery a few years later quickly brought them back to life. This, and most of the important discoveries of the period, centered about one person—Ernest Rutherford. Rutherford was, without a doubt, the towering figure of the early 20th century. Born in New Zealand, he developed an early interest in physics, but soon realized that if he was ever going to be successful he would have to go to England. So he applied for a scholarship to Cambridge. But—alas—he came in second, and only one scholarship was to be awarded. Disappointed, he returned to his parents' farm in the country. He was hoeing potatoes one day when a letter came stating that the winner had declined the scholarship and it would be awarded to him. Overjoyed, he flung down the hoe and yelled, "That's the last potato I'll ever hoe." And indeed it probably was; even in later life he never returned to gardening.

Oddly enough, Rutherford was not among those dissatisfied with Thomson's model. Sure, it had its shortcomings, but Rutherford was nevertheless convinced it was basically correct. But over the years, first under J. J. Thomson at Cavendish, and for a period of nine years at McGill University in Montreal, Canada, Rutherford had developed, honed, and refined various scientific techniques. Among his "tools" was what might be thought of as the first particle accelerator. He was able to direct a beam of alpha particles (helium nuclei) from a radioactive source toward a target, usually some metal.

He was at Manchester University when one of the pro-

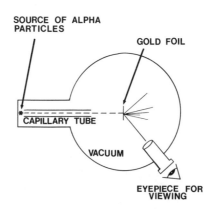

The Rutherford–Marsden experimental apparatus.

fessors working under him, Hans Geiger, mentioned that a student by the name of Marsden was in need of a thesis project. Rutherford suggested that he bombard thin sheets of gold with alpha particles and look for large deflections. He was sure no such deflections would occur; the only thing that would deflect the alphas would be the electrons (assuming Thomson's model was correct), and they were several thousand times lighter than the alphas. It would be like shooting a cannonball at a swarm of bees.

But to his amazement some of the alpha particles did come flying back. Not many, but the fact that any did astounded him. His first thought no doubt was, "It's impossible." How could a bee deflect a cannonball? There obviously had to be something comparable in mass to the alpha particle, something thousands of times heavier than the electron, inside the atom. Rutherford referred to this "something" as a nucleus.

The discovery marked the end of the Thomson plum pudding model. It was soon replaced by Rutherford's "nuclear atom" model in which there was a heavy nucleus at the center, with electrons in orbit around it. How the electrons would stay in orbit was not yet known; the problem of their "radiation death" still hadn't been solved. But Rutherford left details of this type to theoreticians; he dealt in experimental facts, and as far as he was concerned it was an experimental fact that the atom was composed of a heavy nucleus and electrons. Others could clean up the theory.

Incidentally, it might seem that Rutherford would have received the Nobel prize for this work. But he had already received it several years earlier—one of the few cases where a prize winner did his most important work after receiving the prize. Also, to Rutherford's dismay, he wasn't awarded the Nobel prize in physics, but in chemistry. He had mixed feelings about being classified as a chemist, as he had always looked down on them.

To continue the story we must go back a few years to the year 1900. A German professor, Max Planck, at the University of

Max Planck.

Berlin was struggling with one of the important problems of the day. It had been shown that when a metal was heated it emitted more radiation at certain frequencies than others. In fact, for an idealized body called a "black body" the curve at a given temperature was always the same. When the brightness of the emitted light was plotted against frequency (at a given temperature) the curve resembled the hump of a camel; it rose as you went to higher frequencies, then fell off (see figure). Some of the leading scientists of Europe had tried to explain the curve but had failed. No one was able to derive a mathematical expression that "followed the hump."

After struggling with the problem for a while Planck decided to take an entirely different approach. He would "force fit" the curve in an effort to find out what kind of a curve would work, then, once he had the curve, he would explain the physics behind it. With a little trial and error, and perhaps a little luck, he got a mathematical expression for the curve, but the physical interpretation didn't come as easily. He found that he was forced to introduce the idea that the radiation was emitted in "chunks," or what he called "quanta." Furthermore he was forced to introduce a strange new constant he called h. Everything seemed to fit together nicely, but it was revolutionary—so revolutionary that few of his colleagues accepted it. Indeed, Planck himself was by no means convinced it was the final answer.

Yet, strangely, it worked. Furthermore, within a few years Einstein used the idea to explain an effect called the Photoelectric Effect. When light was shone on the cathode of a phototube it produced electrons, which were attracted to the anode. Below a certain frequency no electrons would be emitted; above this frequency their energy was a straight-line function of the frequency. By assuming the light was made up of "quanta" Einstein was able to explain the effect in a remarkable way. We now refer to such quanta as photons. This, in itself, caused a problem, though, since light had a few years earlier been shown to be a wave.

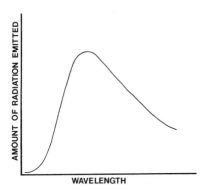

A black body curve. The curve represents the amount of light emitted at various frequencies.

Niels Bohr.

BOHR AND THE NUCLEAR ATOM

Rutherford had laid the foundation; he had given us a simple (but flawed) model of the atom. Furthermore, Planck had introduced the idea of "quanta." What was needed now was someone to bring these ideas together. And, indeed, they were soon brought together by the Danish physicist Niels Bohr.

Shortly after receiving his doctorate in 1912 Bohr came to Cavendish Laboratory at Cambridge. He came with his head in the clouds, sure he was going to set the world on fire. But why wouldn't he—the room where he had defended his doctoral thesis back in Copenhagen was jam-packed with friends, well-wishers, and just plain townspeople, all cheering him on. The newspapers even carried articles about what a great future lay ahead of him.

Confident as he was, Bohr had gone to the trouble of translating his thesis into English and presenting it to J. J. Thomson upon his arrival. He was sure Thomson would encourage him to publish it as the thesis was devoted to Thomson's model of the atom. But no such luck. It lay unread on JJ's desk. In retrospect, though, this may have been in Bohr's best interest as it contained several criticisms of the model. Furthermore, it pointed out a few serious errors.

Bohr was not happy at Cavendish. He had expected much more. The atmosphere was stuffy, there were too many rules, and even worse, nobody knew much about the new "quanta" and nobody seemed to care. The excitement of the continent just wasn't there. Then one day Rutherford came over from Manchester for a visit with his old colleagues and was introduced to Bohr. Theorists weren't among Rutherford's favorite people, and he had even less use for those from the continent, but he took an instant liking to Bohr. As an avid football fan, he was pleased to find that one of the stars of the Danish Olympic team was Niels's brother.

Bohr decided then and there that he had had enough of Cavendish; an invitation to visit Manchester was extended and he quickly took advantage of it. And he was soon glad of his decision. The atmosphere at Manchester was much more to his liking—ideas flowed freely and there was an air of free discussion. Besides, Rutherford was flattered by his interest in the nuclear atomic model (most of Rutherford's contemporaries had expressed skepticism).

Bohr stayed at Manchester for only about six months, but it was an important six months. Ideas that were to blossom later were already forming. Planetary models were spinning around in his head. "Why didn't the electron spiral into the nucleus?" he asked himself over and over. According to Maxwell's theory they were supposed to. There had to be something else—an important basic principle—at work. There were problems, but the planetary model had so much going for it. Bohr was sure that if Planck's quanta could be combined with it a revolution in physics might occur.

When he returned to Copenhagen he continued working

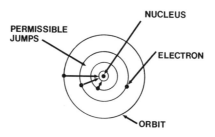

Bohr's atomic model.

on the troubled model. He convinced himself that radiation was only emitted when the electron jumped between certain allowable orbits. But how could he prove it? He was still mulling over the problem when an old classmate visited him in Copenhagen. The friend asked him casually if his model explained the Balmer series. Bohr looked at him, puzzled; he had never heard of the Balmer series. His friend explained that they were hydrogen spectral lines. Bohr quickly looked them up, and within minutes saw that this was what he was looking for: a series of spectral lines with a mathematical formula giving their position. All he had to do was incorporate this formula into his ideas. And within a short time he had a model of the hydrogen atom that completely explained the Balmer series. In Bohr's atom the electrons were in specific orbits according to their angular momentum (mass times velocity times distance from center to orbit). Those with lower angular momentum were close to the nucleus; those with higher angular momentum were farther out.

The idea was radical, however, and few rushed to embrace it. Still, they couldn't argue with it: where it really counted Bohr had hit the nail on the head. He could predict to a high degree of accuracy the position of each of the spectral lines, using nothing more than basic constants of nature (including, of course, Planck's constant, h).

But strangely, when Bohr tried to extend his theory to the

helium atom he found it didn't work. Something was wrong. He struggled for years trying to modify the theory, trying to make it succeed, but he failed. Had he just been lucky? We'll see later that a sweeping new idea was needed, one even more revolutionary than Bohr's. But Bohr had laid the groundwork; he had shown that the introduction of quanta was the key.

Bohr's stay at Cambridge was not a helpful or satisfying one as far as he was concerned, yet there was something about Cavendish he admired: it was a national laboratory of considerable prestige where scientists could meet, talk, and work with other scientists in their own field. Denmark had no such facility and Bohr decided soon after his return to Copenhagen that it needed one. What he wanted was an institution where Europe's best could come to talk about their latest theories. And he managed to persuade a most unlikely candidate—the Carlsberg brewery—to support it. Scientists from throughout Europe flocked to the new institute and it soon became an international center of theoretical physics. Many of the ideas of the next great revolution in physics blossomed here.

The first of these ideas, however, came not from Denmark but from France in 1922. Louie de Broglie, a French prince, became fascinated with Einstein's theories. He began to wonder if there was any relation between Einstein's most famous theory, relativity, and the theory of light. But he made little progress, mostly because he was ill-equipped to handle difficult physics problems; his degree was in history, not physics. Then came World War I and his study was interrupted. But his interest in physics did not wane; in fact, it increased as a result of working with radio communications. When the war was over he therefore decided, with a little encouragement from his brother, a well-known experimentalist, to get a doctorate in physics.

While working on his degree he became convinced that there was a connection between the Bohr atom and the wave–particle duality of light. If a photon could act like a wave sometimes and a particle at other times, why couldn't an electron also

Louie de Broglie.

act like a wave? Making use of Einstein's early work he derived a relationship for the wavelength of such a wave. He then used it to explain Bohr's quantization condition.

To understand what he did, consider some of the properties of waves. What we are interested in, in particular, are standing waves. You can easily see these waves by tying one end of a rope to a doorknob, standing back with the other end in your hand and shaking it. If the rope is of the right length and you shake it at the proper rate you will get a single loop between the knob and your hand. If you shake the rope at a higher rate it is possible to get two or even more loops in the length. de Broglie used the formula he derived to calculate how many such loops there were along the various Bohr orbits (see figure). And interestingly he found there was always an integral number. This meant that the orbits were created on the basis of the wavelength of the wave associated with the electron.

de Broglie waves around a Bohr atom.

de Broglie wrote up his results in his doctoral thesis in 1924 and presented them to his committee. He was still uncertain of the theory at this stage and couldn't explain why the electrons acted like both a particle and a wave. His only justification was that the scheme worked. His committee was skeptical but kept a cool head. Paul Langevin, de Broglie's thesis adviser, decided to ask Einstein's opinion. And to his amazement Einstein was fascinated. The following year Davisson, in the United States, showed experimentally that the idea was correct. Needless to say the thesis was accepted and it eventually became one of the most famous theses in physics—the only one to ever win its author a Nobel prize.

THE DISCOVERY OF QUANTUM MECHANICS

In the early 1920s, as today, much of the theoretical research centered around colloquiums attended by only a few individuals. Ideas were bounced back and forth and argued about. Such colloquiums took place at both Zurich, in Switzerland, and Göttingen, in Germany. And it can be said that the quantum theory of the atom—quantum mechanics—was, for the most part, born through discussions in these cities.

Göttingen was, perhaps, an unlikely place for the birth of such important ideas. There were larger universities, but none more prestigious in physics during the 1920s. The most important event, however, the discovery of the equation that became

Erwin Schrödinger.

the centerpiece of quantum mechanics, took place not in Göttingen, but in Zurich. There were two institutes in Zurich: the Swiss Federal Institute of Technology and the University of Zurich. There were too few physicists at either institute for a good colloquium so they joined forces. Faculty from both institutes attended and gave lectures. The colloquium was led by Peter Debye.

At the end of one particular colloquium Debye looked around for someone to give the next. News of de Broglie's startling discoveries had leaked out but most of those present knew little of the details. Debye's eyes finally stopped on Erwin Schrödinger, a professor from the University of Zurich. "Schrödinger," he said, "you're not doing anything important now. Why don't you talk about de Broglie's work?" Schrödinger nodded, indicating he would.

Schrödinger had received his Ph.D. from the University of Vienna in 1910; he had also served as an artillery officer in World War I. But his career had gone nowhere, so at the end of the war he decided to chuck physics for philosophy. When he tried to get a university position in philosophy, however, he encountered difficulties so he reluctantly remained a physicist. Part of the problem, perhaps, was that he didn't get along well with others and preferred to work alone. The English physicist Paul Dirac once remarked, "with his rucksack on his back Schrödinger looks like a vagabond." And indeed on one occasion he tried to check in at a large hotel but was stopped because they didn't believe he was a professor.

At the following colloquium Schrödinger gave a clear account of de Broglie's work, describing his standing waves and how they accounted for Bohr's orbits. When he was finished Debye stood up and said, "Schrödinger, you are talking foolishly. You are talking of waves, but you have no wave equation." A seemingly unimportant remark perhaps, but it caused Schrödinger to begin thinking about how such an equation could be set up. And a few weeks later when he presented a second lecture on the subject he began with, "Professor Debye suggested that one should have a wave equation. Well, I have found one." And it was soon evident that Schrödinger had made a major discovery. He knew that others would jump on the bandwagon and take advantage of his new technique so he quickly worked out most of the important applications of the theory. Paper after paper appeared with Schrödinger's name on it. In fact, his series of papers were eventually collected into a book, a book that is sometimes used even now as an introduction to quantum mechanics.

Schrödinger's equation was, without a doubt, a major breakthrough. Yet there was something strange about it: it described a mysterious wavefunction he called psi. And nobody, including Schrödinger, knew exactly what psi was. It might seem odd that you could make calculations and derive physical results without knowing what part of the equation meant, but it turns out that this is possible with differential equations of this type.

To Schrödinger, psi was a wave representation of the electron. He thought of the electron as a kind of superposition of waves—a "wave packet." The major difficulty with this view was that the wave packet, though initially small and compact, eventually got fatter and fatter. Real electrons didn't do this. Max Born, at Göttingen, found it difficult to accept Schrödinger's wave packet interpretation. He could see the path of an electron in a chamber specially designed to display tracks of particles (called a Wilson cloud chamber after its discoverer), and it didn't expand with time. The more he thought about it, the more he became convinced that psi didn't represent the electron itself. In 1926 he proposed that it was a "probability" wave. In other words, psi (or more exactly, the square of psi) gave only the probability of finding an electron at a given point. With this, quantum theory took on a new twist. It was no longer a theory in which the properties of the atom could be determined with certainty; only probabilities could be calculated. The electrons in atoms were "probability clouds." In other words the orbits were not definite: the electron could be in many different positions near the maximum of its orbit, and if a large number of measurements were taken and plotted you would get a smeared-out orbit.

Confirmation of this probability idea came within a short time from the German physicist Werner Heisenberg. At this time Heisenberg was already well known. Even before Schrödinger had presented his wave equation Heisenberg had formulated a different version of quantum theory, one we now refer to as the matrix version.

After receiving his Ph.D. at the University of Munich in 1923 Heisenberg came to Göttingen to work with Max Born. Born introduced him to the new idea of the quantum. In 1924 he worked with Bohr at Copenhagen. While most scientists preferred to think of the atom as a tiny solar system Heisenberg disliked such pictures. He felt it did little good trying to visualize the electrons in orbit around the nucleus. The important thing, as far as he was concerned, was the numbers that experimentalists produced; for example, the numbers associated with spectral lines. In 1925 he went to a vacation spot on the North Sea for

Werner Heisenberg.

health reasons. With him he had arrays of experimental numbers that spectroscopists had come up with. These arrays are now known as matrices, but Heisenberg had, at the time, never heard of matrices. This was not a problem to him, though; whenever he needed a new mathematical technique he just invented it. Within a couple of days he had a new theory of the atom—one we now refer to as matrix mechanics. But most physicists of the day were not familiar with matrices, and it was not well received. Schrödinger's approach, which came a year later, was much less foreign, and it therefore got a lot more attention. Schrödinger soon showed, though, that the two theories were really just different forms of the same theory.

Heisenberg's confirmation of Born's ideas came with the introduction of his Uncertainty Principle. This principle deals with quantities such as momentum and position. It states that there is a fuzziness associated with nature on the atomic level. More exactly, it says we cannot simultaneously determine the momentum and position of an electron to a high degree of accuracy. We can get one of them accurately, but not the other. It's like looking at two objects in a microscope, one slightly above the other. When you focus in on one, the other goes out of focus. If you focus on the first one the second becomes fuzzy. As we will see later the principle applies not only to momentum and position, but also to energy and time.

Heisenberg was, in many ways, a most unlikely person to be attracted to physics. He was a rowdy teenager, frequently involved in fights with communist sympathizers. He was also an enthusiastic mountain climber and an excellent pianist. Felix Bloch describes an incident in which he listened to Heisenberg practice a Schumann concerto (Bloch lived in an apartment directly below Heisenberg's at the time). When the music stopped Bloch heard a knock at his door and when he opened it Heisenberg was standing there. Bloch invited him in and they talked for a while, not about physics, but about music. Heisenberg told him that Franz Liszt had discovered that some of his scales were not smooth enough so he canceled all his engagements for an entire year while he practiced only those scales. Bloch said that he felt at the time that Heisenberg was saying something about himself. For this was the way Heisenberg was. In his search for physical meaning he was a perfectionist. He had an infallible intuition, a knack for coming up with the right answer. Yet he abhored too much abstractness. Bloch once said to him that he had finally determined what space is. "It is simply a field of linear operators," he said.

Heisenberg shook his head. "Nonsense. Space is blue and birds fly through it." He was implying that too much abstraction does no good.

Heisenberg was always friendly and warm according to Bloch, but during World War II he was a feared man. He was in charge of Hitler's atomic bomb project, and with the secrecy of the war no one knew how much progress the Germans were making. Hitler boasted several times that he was on the verge of

Paul Dirac.

producing a super bomb. But when the war ended it was found that the Germans still had a long way to go.

It might seem that two forms of quantum mechanics was one too many. But of course as we saw earlier, Schrödinger showed that they are equivalent. No sooner had he done this, though than a third form appeared. Paul Dirac, in England, presented a form he referred to as "transformation" theory.

Born in Bristol, England, Dirac started out as an electrical engineer but found it difficult to find a job so he switched to mathematics. But it wasn't in mathematics that he made his mark. He soon turned to mathematical physics and within a short time was working on quantum mechanics. Besides introducing transformation theory he also extended quantum mechanics. Schrödinger's equation was not valid at velocities near that of light. Dirac set out to repair the defect, and he soon had a new equation. Oddly, though, this new equation applied only to the electron; in particular it predicted the spin of the electron. But it also predicted something that puzzled Dirac: a particle like the electron but with a positive charge. Dirac thought at first it had to be the proton, but it had to have the same mass as the electron and it was well known that the proton was almost 2000 times heavier. It obviously couldn't be the proton. In 1930 Dirac took the bold step of predicting that there was an "antiparticle" to the electron, a particle that had the same mass but opposite charge. The idea seemed farfetched and nobody took it too seriously until Carl Anderson, in the United States, discovered the antielectron two years later.

Anderson was using a cloud chamber to measure the energies of electrons produced by cosmic rays. These are "rays" (actually they are particles) that come from outer space. His cloud chamber was equipped with a magnetic field, so that negatively charged particles would curve in one direction as they passed through the chamber and positively charged particles would curve in the opposite direction. The amount of curvature gave a good estimate of the mass of the particle. What Anderson found was a particle with the mass of the electron that curved the wrong way. In essence it acted as if it was a positively charged electron. He named the new particle the positron, a name we still use today. He also suggested that the electron should be called the negatron, but the name never caught on.

We now know that to every particle there is an antiparticle, and when a particle and its antiparticle meet they annihilate one another with the release of considerable energy. This energy can take the form of other particles, or photons (the "particle" associated with light).

This brings us to an important concept called the "virtual photon." To see how such photons can exist let's look back at the Uncertainty Principle. I mentioned that there is a fuzziness associated with energy and time at the atomic level. An important consequence of this is that a particle such as an electron can emit a photon with the provision that it reabsorb it within a very short time—a time that depends on the energy of the photon. In short, a photon can "sneak out" from the electron under the cloak of the Uncertainty Principle as long as it gets back fast enough. Indeed, it doesn't even have to get back to the same electron; if there is another electron nearby it can be absorbed by that electron as long as it does it rapidly enough. Such photons are called virtual photons; we can never see them but we know they exist. In fact, as we will see later, they are responsible for the electromagnetic force between two electrons. Or for that matter between any two charged particles.

DISCOVERY OF THE NEUTRON

In the late 1920s three particles were known: the photon, the electron, and the proton. But already Rutherford had suggested another might exist. He had noticed that the helium atom was neutral, yet according to the weight of its nucleus it should have four protons in it. To balance the charge of the two electrons in orbit, though, it only needed two. He predicted on the basis of this that there was a neutral particle—a neutron—in the nucleus.

Between 1930 and 1932 several physicists came close to discovering this new particle. Among them were the Joliot-Curies, who bombarded beryllium with alpha particles, much in the same way Rutherford had bombarded gold years earlier. Upon examining the products of the bombardment they found that "some sort of radiation" was given off that betrayed its presence by ejecting protons from nearby paraffin. But they made the mistake of not examining this "radiation" closely enough. For in 1932 Chadwick, in England, performed the same experiment, but decided the results could be better explained by assuming that the alpha particles knocked a neutral particle out of the beryllium nucleus. It was this neutral particle that ejected the protons from the paraffin. This was the "neutron" postulated by Rutherford.

The weight of helium now made sense. The nucleus of the helium atom was made up of two protons and two neutrons. Around them in orbit were two electrons. There was, of course, still a problem: how were the neutrons and protons held together? In fact, there were two problems. It was also noticed that a free neutron, in other words one that wasn't in the nucleus of an atom, decayed to a proton and an electron in about 12 minutes. The phenomenon is referred to as beta decay, because a beta particle is released in the process. (This beta particle is, of course, nothing more than just the electron.)

There was something odd about this decay but before I elaborate further I should say something about the decay itself. The "half-life" of neutrons is about 1000 seconds. By this I mean the time for half of the particles to decay. If, for example, we have 500 neutrons, in 1000 seconds only 250 of them would be left. Now for the odd part. In any decay such as this it was expected that energy and momentum would be conserved; this was a well-known principle of physics, even at that time. In short, it says that the energy (momentum) of the particle before a reaction should equal the energy (momentum) after the reaction. But energy and momentum, much to the surprise of all scientists, were not conserved.

The solution to this problem came from the German physicist Wolfgang Pauli in 1930. He postulated that a third particle, one we could not see (later called the neutrino), was also given off in beta decay. This allowed energy and momentum to be conserved. But the neutrino would be a particularly elusive particle as it had no charge or mass. It was, in fact, so elusive it wasn't found for 30 years. Pauli may have even suggested its existence on a whim, for he never wrote it up in a paper; in fact, he didn't even bother to attend the physics conference at which the idea was proposed. In order to attend this conference, which was held at Tübingen, Germany, he would have had to miss a particularly important (as far as he was concerned) dance in Zurich. So he sent a letter to the conference outlining his idea.

It was an interesting suggestion but few took it seriously.

Wolfgang Pauli.

After all, a particle with no mass that could not be seen in a cloud chamber (the trails of neutral particles cannot be seen) was not something that was likely to get physicists excited.

When Pauli presented the idea he was barely 30, but he had already established a sound reputation. At 21 he wrote a book on relativity that is still used today. In the same year he received his Ph.D., then went on to work under Bohr and later Born. There is no question that Pauli was a brilliant theorist, but as an experimentalist . . . well. To say that he was clumsy is an understatement. It was often joked that whenever he entered a laboratory, most of the equipment in the room ceased to function. But, of course, he wasn't an experimenter, he was a theorist. And later in life he became a critic—a severe one. Scientists had to be sure of their ideas before they came within reach of Pauli. He could easily murder several a day. Once when shown a particularly poor paper he said, "It's so bad it's not even wrong."

Anyway, while he danced in Zurich, physicists at Tübingen

The tracks of neutrinos cannot be seen a bubble chamber. But when they collide, the collision products are visible. A neutrino has hit a particle of liquid hydrogen near the top of this photograph.

considered his idea: the release of a neutrino in beta decay. Perhaps I should be a little more explicit here. The way things have turned out, the particle that is released in beta decay is actually the antiparticle of the neutrino—the antineutrino. Yes, even neutral particles have antiparticles (see Chapter 7, page 129, bottom figure).

It took Pauli a couple of years to present his idea in person. This he did at the American Physical Society meeting in California in 1931. Still, most were not convinced, but among those that were was Enrico Fermi, relatively unknown at the time, but soon to become one of the giants of physics. Fermi liked the idea so much he developed an entire theory around it; he was also the one who named it, calling it the "little neutron," or neutrino. The name is, in a sense, not very appropriate in that, except for lack of charge, the neutron has little in common with the neutrino.

If you look at the properties of the neutrino it's easy to see why it took 30 years to detect it. Not only do most neutrinos pass all the way through the earth without interacting with it, but you could line up 10 earths in a row and most would even pass through them. How, then, do we detect it? Well, the best way, perhaps, is to do what Clyde Cowan and Fredrick Reines did when they detected it in 1956: find a good source. They found, to their delight, that millions of neutrinos were given off in nuclear reactors. Their only problem was how to "trap" them. They accomplished this by introducing a hydrogen-rich target; with such a large number passing through the target they managed to get about three collisions per hour—small but enough to detect. Actually, what they saw was the antineutrino, not the neutrino itself, but of course if it existed the neutrino had to also.

With the trapping of the first antineutrino the "case of the elusive neutrino" was solved. But in reality the story of the neutrino is still far from over. A few years later a particle exactly like the electron, only heavier, was discovered. We now refer to it as the muon. And it also had a neutrino. The question that soon trickled into scientists' minds was: Are the electron neutrino and the muon neutrino the same? An experiment was carried out—and alas—they weren't. They were different. And to make things even worse, an even heavier electron was later discovered (now called the tau), and it too was found to have a neutrino—different from both of the earlier ones.

Then came another surprise: someone suggested that neutrinos might not be massless after all. They might have a small mass. In 1980 a Russian team announced that they had measured this mass in an experiment with a heavy version of hydrogen (tritium). The news stunned physicists. But when an attempt to verify the experiment was made there were problems, problems that still have not been ironed out. So we're still not sure whether or not the neutrino has mass.

YUKAWA AND THE NEW PARTICLES

Now that I've introduced the muon, or μ meson as it is also called, let's look at its discovery. When it was discovered in 1936 a number of scientists were not as surprised as they might have been, for a year earlier one of them had predicted such a particle. This prediction was made by H. Yukawa, a Japanese physicist working at Kyoto University. He did not yet have his doctorate (that wasn't to come for several years) but in 1932 he was already immersed in the major physics problems of the day. The discovery of the neutron and positron had a considerable influence on him, but what influenced him most was a series of papers by Heisenberg in which he proposed that the nucleus consists only of neutrons and protons. Prior to this it was believed that there might also be electrons in the nucleus. In these papers Heisenberg also talked about a force that held the nucleus together—in other words held the protons and neutrons together. He referred to it as an exchange force, but gave no details as to its makeup.

Yukawa thought about this exchange force. He knew that

Hideki Yukawa.

an electron and a proton were held together by an exchange force. The electromagnetic force between them was looked upon as resulting from the exchange of photons between them. Why not apply the same idea to the nucleus, he thought. The exchange particle would have to be different because the nuclear force was quite different from the electromagnetic force, but the idea should work. Yukawa decided that a new particle was involved, one with much greater mass than the electron, one that scientists had not yet discovered. He made the appropriate calculations and showed that the mass of the new particle should be about 200 times that of the electron, an intermediate-mass particle (a tenth as heavy as the proton). He also decided there should be both negative and positive varieties. The nucleus would be held together by the exchange (passing back and forth) of these particles; he referred to them as the U particles. He told several of his colleagues about his discovery. They agreed it was an excellent idea; one of them also pointed out that such particles should be visible in the cloud chamber. But, of course, no such particle had yet been seen. This did not deter Yukawa; he published his idea in 1935.

A year later a particle with a mass of approximately 200 electron masses was discovered. Yukawa's prediction had been verified—or had it? When checks were made it was found that the new particle didn't interact with the nucleus. But Yukawa's particle was a nuclear exchange particle—it had to react with it. Obviously the new particle was not the one Yukawa had predicted after all. But this isn't the end of the story.

Carl Anderson, the discoverer of the new particle, was able to make the discovery because he did something different from what most scientists working with cloud chambers did: he put a lead plate in the center of his chamber to slow the particles down. Most particles passing through a cloud chamber are not greatly affected by a magnetic field because they are traveling too fast. But when they strike the lead plate they are slowed down, and as a result the magnetic field curves them more.

Cosmic rays, coming from outer space as they do, collide with the molecules of our atmosphere almost as soon as they enter it. Because of this it is important to get as high in the atmosphere as possible. Anderson therefore took his cloud chamber to Pike's Peak in Colorado. Within a short time he noticed a track that was less curved than an electron's track, but more curved than a proton's. It had to be a charged particle with a mass between that of an electron and a proton. In fact, there were two such particles, one of each charge. Anderson called the particles mesotrons, but this term was soon shortened to meson.

Another peculiarity of the mesons was their short life. Like the free neutron they decayed, but they decayed much faster—in about 10^{-6} second. They decayed to an electron and two neutrinos. Considerable effort was made to get these new particles to interact in some way with the nucleus. But no dice. There was no question about it: they were not the nuclear glue that held the nucleons (particles of the nucleus) together. Scientists

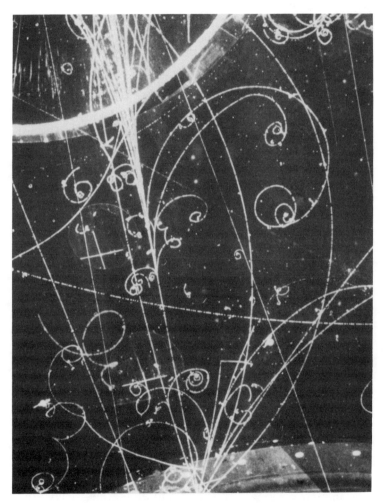

Charged particles in a magnetic field. Note the spiral shape of the tracks.

puzzled over this for years and eventually began to wonder if perhaps there was another medium-weight exchange particle they had not yet discovered. If this new particle did interact with the nucleus you would have to get up high in the atmosphere to see it—before it had time to react with the nuclei of the air molecules. With this in mind Cecil Powell and his group from the University of Bristol in England made a trip high into the French Alps.

Until now most particles had been observed in Wilson cloud chambers, but chambers were bulky, and if many particle tracks were to be seen long stays in the mountains were required. A new photographic technique, however, had been developed and Powell took advantage of it. A stack of specially prepared plates could be set up, left, then picked up later and developed. The tracks of any particles that had passed through them during the time they were set up would be visible on the developed plates.

As expected Powell and his group saw a large number of muon tracks but they also saw some that corresponded to particles slightly heavier than muons. These new particles, later called pions, had a mass about 270 times the mass of the electron, and they also appeared to come in two varieties: positive and negative (later a neutral one was also found). A brief study of the plates then showed something exciting: they interacted with the nucleus. There was no doubt this time: they were Yukawa's particles.

Again, as in the case of the muons, the pions decayed rapidly—in about 10^{-8} second. There now appeared to be two classes of mesons: the pions and the muons. There was no question what the pions' role was, but where did the muons fit in? The physicist I. Rabi, in fact, when first told of the muons screwed up his face and said, "Who needs them?" The muons are, in fact, no longer considered to be mesons. They are the heavy electrons we talked about earlier. We're still not sure what they are needed for, but as we will see later, they do fit into the natural scheme of things—by this I mean they give the universe a certain symmetry.

Despite the apparent sudden poliferation of particles this was really just the beginning. In late 1947 experimenters began finding strange V-shaped tracks in their cloud chambers. Obviously an uncharged particle, invisible in the cloud chamber, was decaying to two particles that were giving the V-shaped track. These two particles were eventually shown to be a proton and a positive pion. The neutral particle was named lambda-zero. It was strange in several respects, but most strange in its lifetime. It decayed after about 10^{-10} second, which may seem like a particularly short time—but it wasn't. Once scientists knew what it decayed to, they knew it decayed via the strong interactions. And the usual time for decays of this type was about 10^{-23} second. Something was obviously wrong; the lambda-zero was living too long.

And there was more. Other V's showed that another family of mesons existed; they were called kaons. And again they decayed too fast. Later, more such particles were discovered; eventually all of them were lumped together and called "strange particles." It was also soon discovered that they were always produced in pairs. When a lambda was produced, for example, a kaon was produced along with it.

The flood of new particles had begun. It was still a trickle at this stage with only the electron, proton, neutron, photon, neutrino, a few mesons, and some strange particles known. Then came particle accelerators.

CHAPTER 3

Particle Accelerators

LAWRENCE AND THE ACCELERATOR

Until the 1930s most of the new particles were detected in cosmic rays using cloud chambers or photographic emulsions. They were generated when energetic particles from outer space, mostly protons, collided with the particles of our upper atmosphere. But we had no control over these collisions. And it soon became evident that if further progress was going to be made we would have to produce our own energetic particles. To do this we would need accelerators.

The race was then on to build a particle accelerator. The first to accomplish the task were John Cockcroft and Ernest Walton, in England. In 1929 they devised a voltage multiplier that built up extremely high voltages and allowed them to attain relatively high accelerations. But in the end their instrument proved to be extremely limited. Nevertheless, they made good use of it, using it in 1932 to bombard lithium, producing alpha particles. What they did, in effect, was combine lithium and hydrogen to form helium—thereby producing the first artificial nuclear reaction. In 1951 they were awarded the Nobel prize for their feat.

But two years after Cockcroft and Walton completed their accelerator, another, much better instrument was invented in the United States by Robert Van de Graaff of MIT. Van de Graaff worked out the principle of the device by using tin cans, silk ribbons, and a small motor, but it worked magnificently; he was able to produce huge potentials. Instruments of this type are still used today at many universities. They are the so-called "lightning makers" you may have seen at science fairs.

But the instrument that was to overshadow them all was the cyclotron. Ernest Lawrence, at Berkeley in California, became convinced that there was a major problem with the accelerators of the day. Particles were being accelerated with a single gigantic push (as a result of a large electrical potential). Lawrence became convinced there was a better way: a lot of small pushes would give the same end result. In fact, if continued long enough such a machine could easily outdo the larger single-push machines.

Lawrence was born in South Dakota in 1905. He got his Ph.D. at the age of 20 from Yale, taking a job at Berkeley in 1928. Within two years he was a full professor—the youngest on the faculty. It might seem on the basis of this that things came easy to him. But they didn't. Everything he attained—and he attained a lot in his lifetime—came as a result of a tremendous amount of hard work. As a university student he supported himself by selling kitchen items to farm wives. Eighteen- and twenty-hour workdays were common to him, but in the end, perhaps, he paid for it. He died at the relatively young age of 57—from overwork. Yet in his short lifetime he achieved much more than most who had lived almost twice as long.

Lawrence was in the Berkeley library one night going through foreign journals when he came across a diagram that caught his attention. The article was in German, and his German was marginal, but he could easily understand the figures. He quickly scanned the article reading as much of it as he could, and it was soon evident that this was exactly what he was looking for. The article described a simple device that accelerated particles by giving them "little pushes." Lawrence's mind raced ahead as he looked at the figures; he could see how the device could easily be improved. A magnetic field could be used to curve the particles into circular orbits. As they passed a particular point in their orbit they would be given a boost. This would take them to a larger orbit. Then again as they passed the same point they could be given another boost.

The next morning Lawrence raced to his lab and along with former student Stanley Livingston, began constructing a simple test model. There were a lot of bugs to get out before they would have a working model, but fear of failure was not one of Lawrence's traits. He was certain it would work, and when they came across a seemingly insurmountable problem he would just work twice as hard to overcome it. Most of the instrumentation and devices needed were not available commercially so Lawrence invented and built his own.

Finally it was complete. It was small—only a few inches across—a small pillbox affair, but it worked. It was the first of a long line of cyclotrons that Lawrence was to build. It consisted of two dees (so named because they resembled the letter D—see figure p. 53). Protons were introduced near the center; a voltage was applied between the two dees so that the protons would be accelerated toward the negatively charge dee. When it reached that dee the voltage direction was reversed and it was accelerated back to the first dee again. Its overall path inside the dees was a spiral. Finally, at the outer edge of the device it reached its maximum acceleration and could be released from the magnetic field in the form of a linear beam.

From the 4-inch accelerator Lawrence and Livingston went on to an 11-inch model. It was with this model that they produced one million volts (1 MeV). From the 11-inch model Lawrence went to a 27-inch one. The instruments were now getting too large for the physics building so Lawrence moved to an adjacent wooden structure. It was the first "Lawrence Radiation Lab," and should, perhaps, have been preserved as a historical monument. Unfortunately, in 1959 it was torn down to make room for a more modern building.

The feats accomplished in the "wooden" lab were phenomenal. Lawrence's staff operated around the clock, seven days a week. Part of Lawrence's success was, no doubt, due to his tremendous enthusiasm—an infectious enthusiasm. Students

Ernest Lawrence and his 4-inch cyclotron. (Courtesy Lawrence Berkeley Labs.)

and others who worked for him thought nothing of putting in 80 or more hours a week. In fact, if you only worked 70 hours a week you were looked down on as not really interested in physics. But Lawrence worked just as hard as anyone; eventually, though, he had to leave more and more of the work to his coworkers. A more pressing problem arose: money. If he was to build a larger cyclotron he would need money—big money. And he went after it. Selling the cyclotron to large grant-awarding foundations became a vital part of his workload. But even here he excelled; he seemed to have a knack for convincing people that the money would be well-spent.

The principle of the cyclotron. Charged particles are attracted to the dee with the opposite charge. The charge on the dees in then reversed. Overall orbit is a spiral.

By 1937 Lawrence had pushed the cyclotron to 8 MeV. Important experiments were now being performed. It was being used in "atom smashing" experiments, and a considerable amount about the nucleus was being learned. Lawrence also became interested in its application to medicine. Along with his brother John (an M.D.) he used it to produce radium and other radioactive elements that could be used in the treatment of cancer. Furthermore, he showed that the particles from the machine itself could be used to treat tumors.

His interest in this area paid off in an unexpected way. In 1938 his mother was diagnosed as having terminal cancer. He rushed her out to Berkeley and had her treated with radiation—one of the first such patients. She lived to the age of 83.

Luis Alvarez was sitting talking to Lawrence one Sunday

One of Lawrence's early cyclotrons—a 60-inch instrument (1939). (Courtesy Lawrence Berkeley Labs.)

afternoon when Lawrence's son came in. "I've been asked to explain the cyclotron to my class," he said to his father. Lawrence pulled up a pencil and pad and began explaining it to him. "The particles go around in circles, curved by magnets, and when they go faster the magnets can't bend them so easily so they go to bigger orbits. Particularly important," he pointed out, "is the fact that the particles in little circles take the same time to go around as those in big circles. This means the ones in big circles go a lot faster." When Lawrence had finished his son looked at him and said, "Gee, Dad, that's neat." Alvarez said that he felt the Nobel prize committee must have had that same feeling when they voted for him. He received the Nobel prize in 1939.

Lawrence continued building larger and larger cyclotrons;

Lawrence Berkeley Labs' 184-inch cyclotron during WWII. (Courtesy Lawrence Berkeley Labs.)

he hoped to get one that would produce 100 MeV, but it turned out that there was a limit. The device would not work over about 25 MeV. If we look at the energies needed in the study of elementary particles, though, we find this is relatively low. A method for obtaining higher energies was obviously needed, and it soon came. In the cyclotron the magnetic field is constant and particles are introduced at the center. As they accelerate in the dees they move to larger orbits. Let's suppose, though, that we accelerate the particle with a cyclotron to a relatively high energy, then introduce it into a hollow circular tube surrounded by magnets. When we increase the energy of the particle again, it, as expected, moves to an orbit of greater radius, but if we increase the magnetic field at the same time we can keep the particle in the same orbit. What we are doing, in essence, is

An aerial view of Lawrence Labs today. (Courtesy Lawrence Berkeley Labs.)

compensating for the increased acceleration by increasing the magnetic field. Instruments of this type are called synchrotrons, and most of the world's largest are of this type. Note that the beam in this case would consist of pulses of particles, since a group of particles would have to be taken up to maximum energy. In the cyclotron we get a continuous beam.

If we try to accelerate electrons in such an instrument, though, we soon find we are in trouble. All particles radiate part of their energy away when they are accelerated, and the amount that is radiated away depends on the mass. Protons, being relatively heavy, do not radiate much; electrons, on the other hand, are light, and radiate a lot. In the case of the electron, after about ten billion electron volts, it's a losing battle—all the energy you give them is radiated away. To get around this problem scientists have built linear accelerators. In these accelerators electrons are accelerated in straight lines and as a result their radiation loss is considerably less. SLAC, the largest accelerator of this type, is at Stanford University in California. As will be seen in later chapters a particularly important series of experiments were performed with this accelerator in the late 1960s.

DETECTORS

Once we have accelerated particles to exceedingly high energies and directed them at a target, it is obviously important to see what happens. In the case of cosmic rays scientists were able to observe the tracks of particles using a Wilson cloud chamber. It contained gas in a rarefied form, and when a particle passed through it electrons were knocked from atoms, creating ions. Water droplets formed along the path, making the track visible. But the number of droplets that formed was relatively small and rare events and particles with particularly short lifetimes could not usually be seen.

Don Glaser of the University of Michigan turned his attention to this problem in the early 1950s. He was sitting having a beer one evening, watching the bubbles rise from the bottom, when a thought struck him: could you cause bubbles to form along the ion path of a particle? As Lawrence did with his cyclotron he immediately began constructing a small test model. It was only a few inches across. He filled the chamber with ether (later he switched to liquid hydrogen) near its boiling point and put it under pressure using a piston. When a particle passed through the chamber he quickly pulled out the piston, lowering the pressure. And sure enough, as expected, bubbles formed along the ion path. And they gave a much more visible track than the cloud chamber.

Large models were soon built. The device became known as the bubble chamber, and for years it was the standard particle detector in accelerator labs. In recent years, though, it has been generally replaced by computer-controlled electronic detectors. Don Glaser received the Nobel prize for his invention in 1960.

Detectors and accelerators are, without a doubt, the two major tools of the particle physicist. The large number of discoveries that have been made in the last few decades are directly a result of them. In the next chapter we will begin our look at these discoveries.

CHAPTER 4

Organizing the Particle Zoo

As we saw in Chapter 2 the major discovery of the early 1950s was the V particle, so called because of the shape of the track they left behind on photographic plates. Actually, if you viewed the track in the right way it was an inverted V. To the scientist of the day the explanation of the V was obvious: a neutral particle (which left no track on the plate) had decayed to two charged particles. Of course they then had to identify the two charged particles. And they soon did. Some of the V's decayed to a proton and a negative pion (they are now called neutral lambda particles) while others decayed to positive and negative pions (they are now called neutral kaons).

The particles were strange; they were an enigma, not because of what they decayed to, but because of their lifetime. They decayed in about 10^{-10} second, which by most standards is an extremely short time. But they appeared to be decaying via the strong interactions (interactions associated with the strong force) and should therefore have lived only 10^{-23} second. Something was drastically wrong—they were living ten trillion times too long. Because of this strange behavior, they were eventually referred to as strange particles.

The behavior of the strange particles created an air of confusion. But out of the confusion came a call for action. Abraham Pais, a physicist who fought in the Dutch underground during World War II, and spent several years in a German concentration camp, was convinced that an understanding of the strange

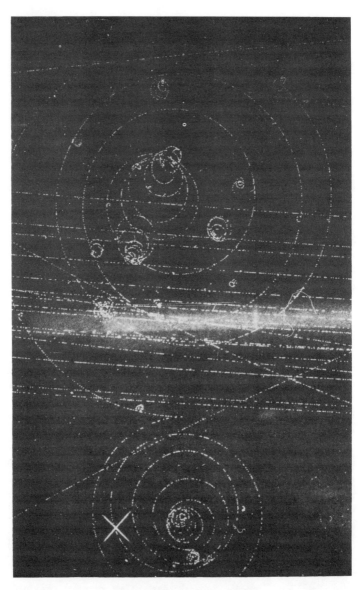

Particle tracks in bubble chamber.

particles was crucial—the key to a much greater understanding of nature. He was certain that they didn't decay via the strong interactions, and this might be the reason for their long life—but he couldn't prove it. He did, however, make a suggestion that eventually led to the proof. He suggested that the strange particles were produced only in pairs, "associated production," as he called it. This meant that when a kaon was produced, a lambda was produced along with it. And indeed, his prediction was borne out.

Still, there were many unanswered questions. Why were they created in pairs? Why was their lifetime so long? Although he was not able to answer these questions himself, Pais was the driving force behind the assault on them.

The discovery of the first strange particle was, in a sense, the beginning of the "particle population explosion," but the explosion didn't really take off until a couple of years later when the first resonance (short-lived particle) was discovered by Enrico Fermi and his colleagues in 1932 at the University of Chicago. Fermi was already world famous as the producer of the first sustained fission reaction. Born and educated in Italy, he was a professor of physics at the University of Rome during the era of Mussolini. Although he was not in extreme danger himself, his wife was Jewish, and as the Nazi influence spread into Italy Fermi feared for her life. It would have been difficult for him to leave Italy, but when it was announced in 1938 that he had won the Nobel prize he saw his chance. He traveled to Sweden with his wife for the ceremonies, then left directly for the United States.

To see how Fermi discovered the resonance we must first look at what is called "cross section." Cross sections play a fundamental role in experiments in which one type of particle is bombarded by another (called scattering experiments). In essence it is the "target area" that the bombarding particle sees. Physicists measure it by counting the number of particles that are scattered out of the bombarding beam. Fermi and his group projected pions at protons and plotted the resulting cross sec-

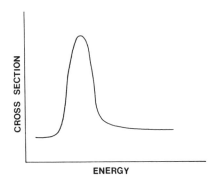

"Bump" in cross section, corresponding to a resonance.

tion for various energies. To their surprise there was a bump in the plot (see figure). Such a bump would occur only if the proton and pion briefly "fused" together. At most energies the pions were apparently just scattering off the protons, but at a particular energy a fusion or binding together was occurring. The lifetime of the fused state could be obtained from the width of the bump, and its mass from the energy of the incoming pions.

Let's take a moment to look at the determination of the mass. Particles, it turns out, can be created in accelerators as long as we have enough energy to produce them. The energy needed is the energy equivalent of their mass. This is, in fact, why we usually talk about the masses of particles in terms of their energy, rather than their weight. We say, for example, that the proton has a mass of 938 MeV. What this means is that it takes 938 million electron volts to create it. Similarly it takes 0.51 MeV to create an electron. In the case of the resonance discovered by Fermi the incident energy was about 1230 MeV; this meant that the resonance had a mass of 1230 MeV.

Can we say that such a short-lived fusion is really a particle? At first physicists were reluctant, and that is why the word "resonance" was used. But eventually they were accepted as short-lived particles. The one discovered by Fermi was called delta (Δ). Later several other deltas were discovered; in all we now have Δ^{++}, Δ^+, Δ^-, and Δ^0, where the sign at the upper right denotes their charge.

If the pion and the proton fused together to create a new particle it seemed reasonable that other such pairs would do the same thing, which proved to be true. Like the delta they were also detected as bumps on cross section plots; so many were found, in fact, that "bump hunting" soon became the fashion. Literally hundreds of new particles were identified in the years that followed.

In a first attempt to make sense out of the chaos the particles were classified according to their weight. The heavier ones, such as the proton and the neutron, were called hadrons, and the light ones, such as the electron, were called leptons. The hadrons were divided into the baryons and the mesons, the mesons being of medium weight. This helped, and strangely, it wasn't all three classes that got out of hand. Only a few leptons are known at the present time; most of the new particles were baryons. But why were there so many of them? How were they related to one another? Something further was obviously needed to understand them.

Physicists turned to the quantum numbers that are associated with particles, hoping they might be the key. (They are designated by the letters Q, M, J, I, B, P.) Each of the particles was found, for example, to have a particular spin. We can think of this spin as similar to that of a top, but we have to be careful in pushing this analogy too far. We'll see later that an electron is believed to be a point particle, in other words a particle with no dimensions. It is difficult to see how such a particle could be physically spinning. Anyway, the concept is useful and it is used with all particles. The electron, for example, is said to have a spin of $1/2$ (more exactly $1/2\, h/2\pi$). Furthermore, it can spin in either a clockwise direction or a counterclockwise direction; we refer to these as spin up and spin down states.

In the mid 1930s Heisenberg introduced a quantity he called

isospin. He noticed that, except for their charge, the proton and the neutron were virtually identical. He therefore introduced the idea that they were different "states" of the same particle, referring to this particle as the "nucleon." He imagined this nucleon to be spinning like a top in an imaginary space called isospin space. If the spin axis was in the up direction the particle was a proton; if it was in the down direction it was a neutron.

Few people paid any attention to Heisenberg's suggestion at first, but eventually it caught on. In 1938 N. Kemmer, a Russian émigré working under Wolfgang Pauli, extended the idea to pions. He suggested that the three pions (π^+, π^-, π^0) were the same particle, each in a different isospin state. The idea was later applied to other groups.

Isospin soon became a valuable classification tool but before we say any more about it let's look briefly at some of the other quantum numbers that are used. As you know, particles also have a charge; associated with it is a quantum number Q that can take on negative or positive integral values, and of course zero (for neutral particles). Another quantum number (M) is associated with the mass or energy of the particle, and one (P) with what we call parity. We will talk about parity in detail later; for now I will only mention that it has to do with whether the mirror image of a particle interaction is possible or not.

Quantum numbers are particularly valuable because they obey conservation laws. You are no doubt familiar with at least one such law—the conservation of energy. It states that the total energy before a reaction has to be equal to the total energy after. In the same way, the sum of the quantum numbers has to have the same value after an interaction as it did before. Such conservation laws are, however, not satisfied by all types of interactions. Isospin, for example, is conserved in the strong interactions, but not in the electromagnetic and weak.

While we are talking about interactions I should briefly review the four major types of interaction fields, as we will encounter each of them sooner or later. You have already been introduced to the electromagnetic and strong fields. The elec-

	Gravitational	Electromagnetic	Strong	Weak
Range	∞	∞	10^{-13} cm	$<10^{-13}$ cm
Exchange particle	Graviton	Photon	Gluon	W^+, W^-, Z^0
Example	Astronomical forces	Atomic forces	Force between quarks	Beta decay
Strength	10^{-39}	$1/137$	1	10^{-3}

tromagnetic force is the one that holds the atom together; the positive charge of the protons in the nucleus attracts the negative charge of the electrons in orbit around it. Within the nucleus there are, however, both protons and neutrons. They are held together by the exceedingly strong, but short-ranged nuclear force. It turns out that there is also a weak nuclear force that is important in the decay of radioactive nuclei, and in, for example, the decay of the neutron (beta decay). The last of the four forces is the one you are likely most familiar with—the gravitational field. It holds you to the earth, and of course it also holds the earth in orbit around the sun.

According to quantum field theory (to be discussed in the next chapter) each of these fields is created by the transfer of a virtual particle. An electron and a proton, for example, are attracted to one another as a result of a transfer of photons. The "exchange" particle in this case is the photon. Similarly, in the case of the strong interactions the exchange particle is a particle called the gluon, and for the weak interactions it is what is called the W particle. Finally, for the gravitational field it is the graviton. Of these four particles we have detected the photon and the W particle. It is assumed, however, that the others exist.

Now back to isospin. For a while it proved invaluable in grouping particles into small families. The neutron and proton, for example, were grouped into an isospin "doublet," with one component corresponding to the proton and one to the neutron. They had the same mass (approximately), the same spin and isospin, but different charge. The pions were put in an isospin triplet, the sigma particles into another triplet. And as more and more particles were discovered they were assigned to other isospin families, or "multiplets" as they were later called.

Then came Murray Gell-Mann, a prodigy who entered Yale at 15, graduated at 19, went on to MIT and obtained his Ph.D. three years later in 1951. After working for a few years under Fermi at the University of Chicago, studying at the Institute for Advanced Study at Princeton, teaching at MIT and in France he eventually settled at Caltech.

In a recent interview Gell-Mann was asked if there was one event that was important in shaping his approach to physics. He answered with an emphatic "yes." "For years," he said, "I had gotten good grades in science without understanding much. I was a machine for taking notes and regurgitating ideas for examinations. All that changed after I attended the Harvard–MIT theoretical seminar during my first year in graduate school." Until then he had believed such seminars were set up to allow graduate students to impress their peers. The moment of truth came during a graduate student's talk about his doctoral research on the boron 10 nucleus. He stated that he had "proven" that the lowest spin state for boron was one. All of a sudden a grubby-looking man with three days' growth of beard sitting next to Gell-Mann got up and said in a strong accent, "Da spin ain't one, it's t'ree. Dey measured it." Gell-Mann said that almost in a flash of inspiration he realized that the scientific enterprise didn't have anything to do with trying to impress the teacher. The most important thing was coming up with numbers that agreed with observation. It was this agreement that physics was all about.

In the same interview Gell-Mann talked about what is needed to be a theoretical physicist. "The tools are simple," he said. "All you need are a pencil, paper, eraser and a good idea." But he went on to say that most ideas are not good ones, and the equations and scribbles that result from looking into them end in the wastebasket.

Gell-Mann had familiarized himself with the grouping of particles into isospin multiplets, but he was convinced something more was needed. Particle physics was still in a state of chaos. He was particularly interested in the strange particles, and the reason for their long life. At the time each of the isospin multiplets was assigned what was called a charge center—a kind of average charge. In the case of the proton–neutron multiplet the proton had a charge +1 and the neutron a charge 0; the charge center was therefore at +½. The strange particles were also put into multiplets—doublets and triplets. The kaons, for example, were put in a triplet with the charge center at zero.

Gell-Mann was giving a talk one day when he stated that the strange particles had an isospin of one, then quickly corrected it to one-half. During the lecture he started thinking of the mistake, wondering if it might be possible. He looked at the consequences of such a change: the charge center would be displaced by one-half. He soon realized that the idea was not only credible, but was an important breakthrough. Introducing a new quantum number that he called the strangeness number S, equal to twice the displacement of the charge center, he was able to explain a slightly altered form of Pais's "associated production." According to Gell-Mann the total strangness in any interaction had to be conserved. He assigned a strangeness 0 to all nonstrange particles (e.g., protons, electrons) and a strangeness +1 to most of the strange particles; a few, though, had strangeness −2. And since strange particles decayed to nonstrange ones, this meant that in any reaction involving "associated production" one of the strange particles had to have $S = -1$ and the other had to have $S = +1$ (i.e., they have to sum to zero). And indeed this was soon shown to be the case. The long lifetimes of the strange particles were also explained. They could now decay via the weak interactions that characteristically had a longer life.

With the introduction of a new quantum number it became apparent that the particles could no longer be grouped into simple isospin multiplets; strangeness also had to be taken into consideration. But there was the underlying mathematics—group theory, as it is known—to deal with. (A group, as you might expect, is what the name implies—a cluster of elements. A simple group is the group of all integers.) When we deal with isospin we use a group referred to as SU(2) (called the unitary group of 2 by 2 arrays). Gell-Mann realized that a larger group would be needed to incorporate strangeness but not being an expert in group theory he wasn't sure how to proceed.

In the meantime other approaches were being tried. Fermi and his student Chen Ning Yang at the University of Chicago considered the possibility that some particles were made out of others. In particular they made the assumption that the three pions were composed of protons and neutrons (and their antiparticles). They showed that the quantum numbers worked out superbly if the positive pion was composed of a proton and an antineutron. Also, they showed that agreement was obtained if the negative pion was made up of a neutron and an antiproton. You might find it strange that an intermediate-mass particle such as a pion could be made from two heavy particles, but Fermi and Yang didn't overlook this point. They assumed that the excess mass was tied up in the binding energy of the two particles together, in other words, the energy that holds them together.

The Fermi–Yang model was the impetus for a similar, more ambitious model by the Japanese theorist Shiochi Sakata in 1956. Sakata hoped to account, not just for the pions, but for all particles. His first step was to add another particle, called lambda, to the proton and neutron that Fermi and Yang had used. He showed, for example, that the positive kaon could be made up of a proton and an antilambda, and the negative kaon from an antiproton and a lambda; he also had triplets in his scheme. Sakata's model was more complete than Fermi's in that it was based on group theory [the group SU(3)]. Shortly after its introduction, it was extended by several other Japanese physicists, but it was eventually shown to be flawed. It accounted reasonably well for the mesons, but stumbled when it came to the baryons.

THE EIGHTFOLD WAY

Gell-Mann was convinced that Sakata was on the right track but he didn't like his tactics. Some of the particles were not in their proper families. Besides, he wanted a complete field theory, rather than just a scheme for grouping particles. With this in mind he began looking at an idea that had been introduced in 1954 by Yang and Mills of the Princeton Institute of Advanced Study. They showed that certain types of field theories had a "flexibility" or arbitrariness about them—a strange, but useful arbitrariness. Gell-Mann wanted something similar to the Yang–Mills theory, but based on a group that would account for all particles. The Yang–Mills theory was based on isospin and had the group SU(2) associated with it. After learning the basics of group theory he came to the realization that if he wanted to incorporate both isospin and strangeness, a larger group was needed, one that contained SU(2). You might think that it would be logical to try SU(3) immediately; it is similar to SU(2) but made up of 3 by 3 arrays of numbers rather than 2 by 2. In the end this turned out to be the group he needed, but it took many months before he realized it.

Within SU(3) were various ways in which the groups could be represented (called representations). The simplest corresponded to three particles; this was the one Sakata had used. His basic particles were the proton, the neutron, and the lambda. Gell-Mann struggled with this approach and got nowhere. He therefore skipped over the simplest, or "fundamental," representation and went to the next higher order. It had eight particles in it. Within a short time he found that he could group the known particles into families of eight in a reasonable way—with everything fitting (see figure). He was pleased as he looked at family after family and everything fell into place.

Yet everything wasn't to his liking. He had originally hoped to obtain a full-fledged field theory using Yang–Mills theory but problems with this approach became so severe he had to be satisfied with a scheme that merely classified the various particles into families (multiplets). These families are plotted below.

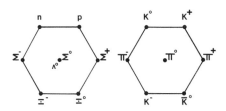

Gell-Mann's families of eight, or eightfold plots. Each of the particles within the family has the same spin, and baryon number, and they all have approximately the same mass.

Gell-Mann referred to the method as the eightfold way. Why? It may sound a little silly, and in fact it is silly, but he wanted to introduce a little humor into an otherwise serious subject. It goes back, I suppose, to his somewhat whimsical personality. Anyway, he had heard that the Buddhists had an "eightfold" path to wisdom within their religion. And with eight particles in his scheme the name seemed appropriate. But—alas—the joke backfired. Others took the name quite seriously and tried to connect his method with Eastern mysticism. Much to his dismay.

There were, unfortunately, problems with the arrays Gell-Mann got using SU(3). First, the families were not complete; in some cases where there should have been a particle there was an empty space. This didn't give him any gray hair (that didn't come until later) as he realized that many particles had not yet been discovered. But if the underlying group theory was to be truly symmetric as it should be, he knew that all the particles within a single family had to have the same mass—and they didn't. To explain this Gell-Mann put forward the idea that the symmetry was "broken." (We'll see later that broken symmetry plays an important role in elementary particle physics.) Using this he and, independently, the Japanese physicist Susumu Okubo were able to derive a formula that gave all the masses of the particles in the families.

So far I've only talked about families of eight, but Gell-Mann noticed that his scheme also gave families of nine and ten.

Murray Gell-Mann.

The mesons fitted into the group of nine quite nicely, but the group of ten was a problem. Gell-Mann therefore neglected it.

While Gell-Mann was working on his eightfold method, half way around the world the same discoveries were being made independently by an Israeli physicist, Yuval Ne'eman. Ne'eman, an Israeli army officer who graduated from the Israeli Institute of Technology in 1945, had an off–on physics career (interrupted each time the Israelis were involved in a war). Finally, though, he decided that the interruptions weren't getting him anywhere so he applied for and got permission to study physics full time. He wanted to get a Ph.D. so he went to England and in London met with and eventually worked under Abdus Salam. Like Gell-Mann he became interested in using group theory to arrange the particles into families with similar properties. And like Gell-Mann he also found he didn't know what to do with the fundamental three-particle representation so he skipped over it to the eight-particle patterns. Salam didn't take Ne'eman's work seriously and didn't encourage him, but when a paper arrived showing that Gell-Mann was working on the same problem and had made some startling progress Salam's interest was perked.

A high point in both Gell-Mann's and Ne'eman's careers came in June 1962 when they attended a conference on elementary particles at CERN. Up to that time neither had paid much attention to the family with ten members in it—the decuplet (see figure), as it is called. Although the four particles that made up the base (the four deltas) were known, and even the line below it (the sigmas), a total of three particles were missing—the three that made up the peak. You can therefore imagine their surprise when an announcement was made at the meeting that two new particles had been found (now referred to as xi-star)—the two that belonged in the next row of the decuplet. This left only one blank in it—the particle at the peak. Since its position in the multiplet was known, literally all its properties were also known: it would be negatively charged, have a spin of $3/2$, and a strangeness of -3. In fact, even its approximate mass was known—1676 MeV. At exactly the same time the two men raised their hands to make the prediction that a new particle, called the

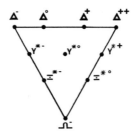

The decuplet.

omega-minus (Ω^-) by Gell-Mann, should exist. The chairman of the meeting looked at both of them, then pointed toward Gell-Mann. Gell-Mann strode to the podium and made the announcement—much to the surprise and dismay of Ne'eman. Later in the meeting Gell-Mann and Ne'eman met for the first time.

Until now most physicists had been skeptical of Gell-Mann's dabbling with group theory. But with the prediction of a new particle there was a surge of interest. Nicholas Samios and Jack Leitner from Brookhaven National Laboratory on Long Island, New York, approached Gell-Mann the moment the meeting was over, asking him for information about the omega-minus. To them the announcement was an "experimentalist's dream," but they were not sure the director of Brookhaven, Maurice Goldhaber, would agree. They didn't want to take any chances so they asked Gell-Mann to write a note to Goldhaber pointing out the urgency of the experiment. Gell-Mann laughed, took a paper napkin from the table where they were sitting and wrote the note. And it worked—Goldhaber agreed to proceed as quickly as possible.

Experimentalists at CERN, where the meeting took place, had already headed for their accelerators. At Brookhaven, Samios and Leitner teamed up with William Fowler and they were soon gearing up for the experiment. The first thing on the agenda was to decide how the omega-minus could be produced. The best way, they decided, was to project a beam of kaons at protons. The mass of the omega-minus was 1676 MeV so they needed that much energy, plus enough to create any other particles that might come out of the reaction. They finally decided a minimum of 3200 MeV would be sufficient, but to be on the safe side they would use 5000 MeV. That was easily within the range of their accelerator.

The experiment began in December 1963, and by the end of January 50,000 bubble-chamber photographs had been taken of kaon–proton interactions. Each of them was examined in detail but nothing indicating the presence of an omega-minus was

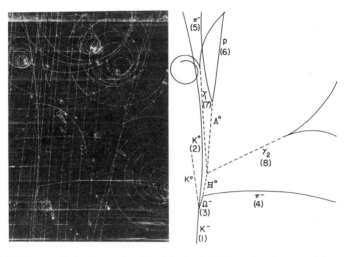

(Left) Photograph showing tracks of particles involved in reaction that created the omega-minus. (Right) Sketch identifying the particles in the reaction. (Courtesy Brookhaven.)

found. Then on January 31, 1964 it came: a set of tracks that could have resulted from the decay of an omega-minus. They had expected that when the negative kaon hit the proton it would produce an omega-minus along with a positive and neutral kaon. The photograph showed the negative kaon interacting and producing a positive kaon but there was no sign of the neutral kaon. This was not a surprise, of course, as it was not expected to leave a track. At some distance above the interaction point was a V-shaped track. One of the two tracks was immediately recognized as produced by a proton; the other was identified as the track of a negative pion. The experimenters were soon able to determine that the proton–pion pair had been created by the decay of a lambda-zero particle (which left no track). Extending back to where the kaon had interacted they found the telltale evidence of the omega-minus. It had decayed to the neutral lambda and a negative kaon. Samios and his group were thrilled. A few weeks later evidence for a second omega-minus was discovered.

So far we've only been talking about the particles themselves: their mass, spin, strangeness, and the families to which they belong. But there's also another aspect of particle physics—the particle interactions. Gell-Mann and Ne'eman had both hoped that their theory would be a complete field theory that would predict the results of such interactions, but it turned out that it wasn't. A field theory was finally developed, however. To see how it was developed we have to go back a few years. We'll do that in the next chapter.

CHAPTER 5

Overcoming Infinity

Shortly after Erwin Schrödinger put forward his equation describing the function psi, physicists became interested in how matter and fields interacted. Michael Faraday had introduced the idea of a field many years earlier, and Clerk Maxwell of Cambridge had put it on a firm footing using mathematics.

What is a field? It can be thought of as a region where a quantity is specified at each point of space. A simple example is a region where, say, the wind velocity is specified at each point. The fields you are likely most familiar with are the magnetic and the gravitational fields. At any point around a magnet (the source of the magnetic field) the intensity and direction of the field can be measured. This also applies to the gravitational field.

With the introduction of quantum theory in the 1920s it became important to be able to treat the electromagnetic field using the new theory. Schrödinger's equation was useful for treating matter, but of limited use when it came to fields. This difficulty was overcome in 1927 when Paul Dirac published a paper showing that both matter and the electromagnetic field could be quantized (put in quantum form). His theory showed, in short, how light could be absorbed and emitted by atoms. It was successful in that it allowed scientists to make many calculations that they hadn't been able to previously. But the theory was limited; it was not valid when the particles had velocities near that of light. In other words, it was not a relativistic theory.

Again Dirac came to the rescue. In 1928 he published an equation that was, in essence, a relativistic extension of Schrödinger's equation. (We now refer to it as Dirac's equation.) It was an important breakthrough. In addition to predicting the spin of the electron, it allowed calculation of the positions of the spectral lines of hydrogen to a high degree of accuracy. But it too had a problem: it predicted both positive and negative energies. What was the significance of the negative energies? No one seemed to know. The known particles certainly didn't have negative energies.

And again Dirac's ingenuity came through, although at the time many thought he had gone off the deep end. He proposed that there was a "sea" of negative energy particles all around us, and that this sea was filled and we could therefore not observe it. This raised eyebrows but with his next suggestion most just shook their head. He said that the positive energy particles could not make transitions to the negative energies because the negative sea was full, but there was the possibility of transitions from the negative sea to positive energies.

What would we observe if this were the case? According to Dirac we would see the appearance of a positive energy electron and the effects of the hole left in the sea of negative energies. This hole would appear as a particle with an opposite charge. But it was here that his tremendous intuition finally began to fail him—he faltered. No particle like the electron, but with a positive charge, was known. Dirac therefore hesitatingly suggested that this particle might be the proton. Oppenheimer soon pointed out, though, that this was impossible; the hydrogen atom wouldn't be stable if the new particle had a mass different from that of the electron. Dirac therefore postulated the existence of an "antielectron," a particle exactly the same as the electron in all respects except charge; it would have a positive charge. If enough energy was supplied to an electron in the negative sea it could be raised to positive energies and in the process produce an electron–antielectron pair.

The idea was preposterous as far as most physicists were

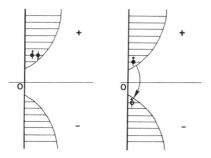

Figure to the left shows filled negative sea. Figure to the right shows creation of hole in the sea.

concerned. But to their surprise a positive electron (usually called a positron) was identified by Carl Anderson a couple of years later. Did this mean that other particles also had antiparticles? According to Dirac they did, but it was a long time before any more were found. The identification of the antiproton didn't come for another 25 years.

Dirac's sea of negative energy electrons was a useful device for its time. But it was only a device. We no longer treat antimatter as having been created in this way.

Dirac's equation was the basis of what we call quantum electrodynamics—the theory of the interactions between electrons and photons. It allowed theorists to make many calculations but eventually it too developed problems. A major difficulty was pointed out in 1930 by Oppenheimer. He calculated the interaction of the electron with its own field, in other words its "self-energy," and found that it was infinite. At first there was disbelief. It didn't seem possible. But Pauli and Heisenberg soon verified it. And although it was discouraging, it didn't come as a total surprise. Physicists had known for years that when they attempted to calculate the same energies without the use of quantum theory (classically) they also got an infinite result if they assumed the radius of the electron was zero. In classical physics, however, the problem had been alleviated by giving the electron a finite radius. But when quantum theory was applied to the problem, giving the electron a finite radius didn't help. The self-energy was now considered to be due to the emission and reabsorption of virtual photons.

Within a short time of the discovery of infinite self-energy another problem arose. It was well known that part of the mass of a charged particle was due to its electromagnetic field. And very close to the electron this field approached infinity. The overall "mass" of the electron was therefore infinite. Theorists had barely adjusted to these problems when another appeared. It seemed as if the charge of the electron would also be infinite. As we saw earlier the electron is surrounded by a cloud of virtual photons, and as you get closer to it the number of these photons grows. Close to the electron some of them are energetic enough to create electron–positron pairs. These pairs exist only for an instant before they come together again to form a photon, but during this instant the electrons are repelled slightly and the positrons attracted slightly. The result is a screening of the "true" charge of the electron. (It is now referred to as vacuum polarization.)

In short, it seemed that we were not seeing the "bare" charge and mass of the electron. It was being screened by a cloud of virtual particles. We got a finite value for these quantities because of this screening. By 1936 it was concluded that the major problem surrounding quantum electrodynamics centered around this vacuum polarization and the infinite mass and

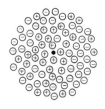

Virtual particles around the electron. The true charge of the electron is screened by these particles.

charge that resulted. How would it be possible to get around this? Victor Weisskopf suggested that we might be able to eliminate the infinities by absorbing them into redefinitions of mass and charge, a technique called renormalization, but it would be difficult to do and would have to be done in a special way. The same suggestion was made again a few years later by H. A. Kramers.

To some, field theory seemed doomed. But others were not worried: things such as self-energies and vacuum polarization were second-order effects, and so far experimentalists had not been able to measure such effects. If they couldn't measure them why worry about calculating them. As early as 1933, though, there was experimental evidence that these effects might have to be taken into consideration. William Houston of Caltech and Y. M. Hsieh from China measured the fine structure of the spectra of hydrogen and found what they believed was a discrepancy. W. E. Williams of London and others also found a discrepancy but there was disagreement among them; furthermore, some argued that there was no discrepancy whatsoever. Oppenheimer and Bohr suggested that the discrepancy could be a result of the neglect of self-energy and vacuum polarization, but because of the controversy little came of the suggestion.

Then came the war and no further work was done on the problem for several years. The war delayed things but it provided Willis Lamb of Columbia University with some invaluable experience. Lamb did his Ph.D. thesis in field theory under Oppenheimer between 1934 and 1938, then went to Columbia University. At Columbia he met I. I. Rabi and became interested in microwave spectroscopy. During the war he became an expert in microwave radar and vacuum tube techniques.

His interest in the energy-level spectrum of hydrogen came about as a result of a class he taught in 1945. The textbook he used referred to the early work by Houston and Williams. In looking over the techniques that they had used Lamb saw that he could do the experiment much better with the new micro-

Willis Lamb.

wave equipment that had been developed during the war. He was sure, in fact, that he could resolve the controversy that had developed. He began thinking about the experiment he would perform. According to Dirac's theory the hydrogen atom could exist in two states (called the $2S_{1/2}$ and $2P_{1/2}$) both of which had the same energy. If microwaves were passed through hydrogen converting the atoms from one of these states to the other, energy would be absorbed if the two states were not exactly at the same energy, he reasoned. He set up the experiment, and after a few false starts managed to show that energy was indeed absorbed. The energy levels were therefore not at the same position; the deviation was small but it was critical, and there was no doubt this time. Lamb was able to measure the effect to many decimal places.

Although a good deal of ingenuity went into the experiment, considerable luck was also involved. Lamb's diverse experience before and during the war was invaluable. He learned of the effect while working as a theoretician under Oppenheimer, but during the war he worked primarily as an experimentalist. His first experiments were failures but he kept at it and eventually met graduate student Robert Retherford who was familiar with many of the techniques needed. With Retherford, Lamb completely redesigned the experiment and was soon able to measure the shift in the energy level of hydrogen. This shift is now referred to as the Lamb shift.

A second important result to come from the experiment was a precise value of what is known as the fine structure constant. This constant, which is a combination of the electronic charge, Planck's constant, and the speed of light, had intrigued physicists for years. Eddington developed an entire (but somewhat dubious) theory around it. Yet nobody knew its exact value. Lamb showed it to be 1/137.0365.

Lamb was informed that he had won the Nobel prize while he was teaching a quantum mechanics class. Unlike most, who would have immediately canceled the class and celebrated, he thanked the messenger and went on teaching the class. Not until it was over did he meet with reporters.

Word of Lamb's experiment soon leaked out, but for many the first inkling of the astonishing result came at Shelter Island in June 1947. The war was now over and physicists had gone back to their old jobs and, of course, their old problems. And the most glaring problem to field theorists was still the infinities in quantum electrodynamics. One of the ways to get things reorganized into a united front against the problem was to call a conference, and in the spring of 1947 one was called. By today's standards it was small, and unlike most of those today that are held at large universities in large cities it was held in a most unlikely place—Shelter Island, a small island off Long Island, New York. Most of the major field theorists were invited: Weisskopf, Bethe, Kramers, Oppenheimer. And along with the well-known old-timers there were several less-known young theorists. Among them were Julian Schwinger and Richard Feynman. Willis Lamb, of course, was also invited.

The conference, although small, was one of the most important ever held. It marked the birth of quantum electrodynamics as a viable and highly accurate theory. A small brass plate commemorating the event can still be seen at the inn where the meeting was held. The meeting opened on the morning of June 2 with Lamb giving the details of his experiment. When he had finished, those present realized that there was now little doubt: higher-order effects in quantum electrodynamics would have to be calculated—and made finite. There was no longer the excuse that they couldn't be measured. Kramers and Weisskopf had already, years earlier, indicated that the second-order infinities could be gotten rid of by "readjusting the constants of the theory," what we now call renormalization. But how could it be done in a logical and consistent way?

Hans Bethe was convinced that a calculation of the self-energy could be made, and on the train ride back to Cornell he began. The full-blown relativistic calculation would be formidable but he decided that the nonrelativistic one would not be too difficult and it would be a useful guide as to whether the renormalization technique worked. Shortly after he got back to Cornell he had completed it. And he was pleased—it worked. It was a crude calculation but it predicted over ninety percent of the deviation observed. He was sure the relativistic calculation would account for the rest. When others heard of Bethe's calculation they were stunned. It hadn't been that difficult after all.

But a more complete and rigorous calculation was needed, and many of those present at the conference immediately set to work on it. Weisskopf had already spent some time with the problem but his progress had been slow. Feynman and Schwinger now threw in their hats, and even Lamb (who had earlier worked as a theorist) began work on it. The first to arrive at a solution was Lamb and a student named N. Kroll. It was an important step, but the method they used for subtracting off the infinities was clumsy and unreliable. A more logical and consistent method was needed—and it soon came.

Julian Schwinger.

The first Shelter Island conference had been so successful a second was organized the next year. This time it was held in Pennsylvania. The same theorists were again present, with the addition of Niels Bohr. The star of the last meeting had been Willis Lamb; this time it was Julian Schwinger.

Schwinger, a prodigy who thought nothing of performing calculations hundreds of pages long, was born in New York City in 1918. It was obvious from an early age that he had a special ability in mathematics. He continually amazed his teachers. He was so far ahead of the other students that he was transferred to Columbia University at the age of 16 without completing high school. In the same year he published his first scientific paper, and a year later he had his bachelors degree. His Ph.D. was completed by the time he was 21.

At college, to the dismay of his teachers, Schwinger rarely attended classes. This almost got him in trouble with the Columbia University mathematics department. They threatened to flunk him if he didn't attend, but a special exam was arranged and he did so well that the threats were immediately forgotten.

Someone once said that by the age of 25 Schwinger knew ninety percent of all of physics—and it would only be a matter of months before he learned the other ten percent. He was, without a doubt, a universalist—perhaps the last. He knew everything.

While he was at college he kept a notebook at his side. Whenever he had some free time it was open and being filled with equations. A professor looked at the notebook and found that it contained the results of numerous scientific papers—important results before they were calculated and published by others.

Schwinger was the first speaker at the Pennsylvania conference. The measurement of the Lamb shift had had a profound effect on him, but it was another result of the conference that got his first attention: an effect related to the magnetic field of the electron, called the anomalous magnetic moment. The calculation, although extremely lengthy and complex, was performed in record time. Then he turned to the Lamb shift and renormalization. Here he literally started at page one. Using the logic and ingenuity he later became famous for, he carefully reconstructed quantum electrodynamics, showing each of the difficulties and how it could be overcome. It was a long and arduous task requiring hundreds of pages of calculations. And he admits making many mistakes along the way, but in the end he tied everything together and renormalized the theory in a logical way.

He was the star of the show. Even Niels Bohr was awed by his tremendous mathematical insight. When he finished the infinities were gone and quantum electrodynamics was a viable theory. Literally every electrodynamical interaction could now be calculated to any order of approximation (at least in theory). And, as was later shown, the agreement with experiment was phenomenal.

But not everyone was as enchanted with Schwinger's meth-

Richard Feynman.

ods as were Bohr and Oppenheimer. They were extremely complex, and the massive mathematical structure he introduced was so unwieldy that only a Schwinger would ever be able to work with it. An irritated critic later said, "Other people publish to show you how to do it, but Julian Schwinger publishes to show you that only he can do it." And, oddly enough, at the same meeting there was a second solution to the problem—a much more easily understood solution. But equally strange it was criticized severely by both Bohr and Oppenheimer. Yet in later years it became the technique that the masses flocked to. This second technique was due to Richard Feynman.

Like Schwinger, Feynman was born in New York City. He

received his bachelors degree from MIT in 1939, and his Ph.D. from Princeton University in 1942. He could easily perform complicated calculations in his head. He liked to bet people that he could work in 60 seconds any problem they could state in 10 seconds. And he usually won. A cursory look at his career (he recently published a funny book outlining much of it) makes you wonder, though, when he had time to do physics. He is a lively person who loves to play tricks on his colleagues, pound the bongo drums at 2:00 A.M., chase Las Vegas show girls, and generally act up. While working in Los Alamos he developed quite a reputation for cracking safes. On occasion he would "borrow" top secret documents from locked safes, leaving a note that he had returned them to the top drawer of a nearby desk. The practice obviously caused some gray hair. Most of the recipients of his jokes did not find them too funny. With all these activities one wonders how he accomplished as much as he did.

But of course he did spend a considerable amount of time doing physics. There were times that he felt burnt out, and times that he got a little disgusted with physics. It is interesting that just before he made what is perhaps his greatest contribution—the renormalization of quantum electrodynamics—he felt his life as a productive scientist was over. He had just helped build the bomb at Los Alamos, and was beginning to have second thoughts about whether he had done the right thing. Furthermore, his wife of a few years had just died of TB. Somehow it all caught up with him and he couldn't do physics; he couldn't even think about physics. He was convinced he was washed up.

Then a miracle happened. To most it certainly wouldn't seem like a miracle but to Feynman it was. He was sitting in the cafeteria at Cornell when a student tossed a plate in the air, giving it a slight spin. Feynman watched it spin and wobble as it rose and fell back to the student's hand. Suddenly he knew what was wrong. He knew why he wasn't getting a "kick" out of physics any longer: he was taking it too seriously. When he was a student he had made calculations, hundreds of them "just for the fun of it." He loved to figure things out, determine how they worked, find out what principle was involved—it was a compulsion with him. But nobody had pushed him then; he had done it because he wanted to. Now he felt an obligation, and as a result the fun had gone. And as a further result nothing was being accomplished.

It was this change of perspective—the idea that he would do calculations for the fun of it, because he wanted to—that caused the miracle. He knew that Hans Bethe was working on the problem of self-energies; he had dabbled with the problem himself some years earlier. He also knew that Bethe had made some progress but he was sure he could make a contribution to the problem. And indeed he did—a contribution that eventually won him the Nobel prize.

Still, even after he made the calculations things weren't a bed of roses. He developed the theory, invented a new and useful diagrammatic technique, and presented the results at the Pennsylvania conference. His talk was immediately after Schwinger's. And where Schwinger's talk centered around a mass of mathematical complexity, Feynman's was centered on something new and strange to those present. He drew little diagrams on the blackboard and used them to explain what he was doing. But nobody could follow him—mostly because so much of what he did was intuitive. Bohr became so exasperated he told Feynman he should go back and learn basic quantum mechanics. (Yet only a few years earlier Bohr had selected Feynman out at Los Alamos as an advisor because he was not afraid to speak his mind.) Feynman had, in fact, never cared for the usual formulation of quantum mechanics and had developed his own approach—now referred to as the path-integral method. Today it is used extensively, but to the participants of the conference it was new—and confusing.

Feynman tried to convince them but finally gave up in disgust. According to Freeman Dyson, who worked with Feynman back at Cornell, he was severely depressed when he got back from the conference. But he was determined—determined to show that his techniques were workable. He would publish them. And, indeed, when he did they were soon accepted; in fact, they were so eagerly accepted that they are now the standard method of making such calculations.

Let's look for a moment at the diagrams he introduced. According to Feynman he began using them out of habit to keep track of the terms in his long and complicated formulas. For the electron he drew a solid line; for the photon a dashed one. As an example let's consider the scattering of two electrons. We know that according to the quantum view they repel one another because they pass virtual photons back and forth. According to Feynman this interaction can be represented as follows:

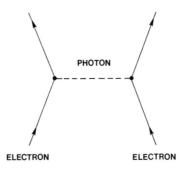

A Feynman diagram depicting the collision of two electrons. A photon is exchanged between them.

On the outsides of the diagram are lines representing the electrons, and between them we see a dashed line representing the photon. The point where the dashed and solid lines join is known as a vertex; this is where the photon is emitted and absorbed, respectively.

Feynman showed that he could represent any of the possible interactions that take place between electrons and photons in terms of such diagrams. Furthermore, he showed that he could use them to write down the mathematical formulas needed for calculating the interactions. His diagrams can also be used to explain renormalization in a simple way.

Before we can talk about renormalization I have to explain what is called perturbation theory—the technique on which quantum electrodynamics is based. It is a process in which the largest contributions to the effect are calculated first (these are the first-order contributions we talked about earlier). But there are also smaller terms that must be added as corrections—second-order terms, and even smaller ones that must be added to them, called third-order terms, and so on (e.g., self-energies). To illustrate further let's look back at the diagram above. Using it we can calculate first-order terms in a straightforward way. It turns out that we can also use similar diagrams to represent higher-order terms. It is, for example, quite possible that the virtual photon could split into an electron–positron pair as it moved between the electrons. In this case we would have a diagram such as

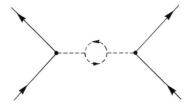

A higher-order Feynman diagram showing electron–positron pair production.

Similarly we could also have

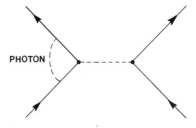

A higher-order Feynman diagram showing emission and reabsorption of a photon.

and a whole host of others, each representing a way the photon can act. These are second-order diagrams and it can be shown that they are supposed to give contributions 1/137 times smaller than the first-order diagrams. But as we saw earlier they don't—without renormalization they give an infinite result. We can see now why the problem was so great—there are obviously a large number of these diagrams. It turns out, though, that with renormalization the contribution from all of these diagrams can be determined at once, and when it is, the theory makes sense. The diagrams give smaller results than the first-order ones, as expected.

Feynman talked about his attack on the problem in his Nobel address. He said he worked on the problem for about eight years (not continuously, of course) before finally solving it. He became familiar with it through books by Dirac and Heitler. His first idea was to make the classical theory (unquantized version) finite, then quantize it. (He worked on this approach while still an undergraduate.) But this turned out to be more difficult than he anticipated. Then came the war and he went to Los Alamos, and little more was accomplished on the problem.

After the war he went to Cornell where he was associated with Hans Bethe. The Lamb measurement and the conference at Shelter Island brought a resurgence of interest in the renormalization of quantum electrodynamics. Bethe gave a talk on his calculations shortly after he returned from Shelter Island. He outlined some of the problems and how they should be solved. Feynman walked up to him after the lecture and said, "I can do those calculations, I'll bring them into you tomorrow." The next day he brought them to Bethe's office and the two of them worked through the problem on the blackboard. But—alas—to Feynman's surprise things didn't work out—the infinities were still there.

Feynman couldn't believe it. He went back to his room and went through them carefully again. And equally surprising—everything was okay. There were no infinities. So back to Bethe's office he went, this time with a successful calculation.

He later said he had never been able to figure out what he had done wrong on the blackboard the first time. From here Feynman refined and developed his methods, finally publishing them in 1947.

Perhaps it was too much to wish for, but surprisingly, a third solution to the problem soon surfaced. When Oppenheimer arrived back at his office from the Pennsylvania conference a manuscript was waiting for him. It was another solution to the infinity problem—apparently different from the other two. It had been devised by Shin'ichirō Tomonaga of Japan as early as 1943. Because Japan was isolated by the war at the time there was no communication between Japanese and American physicists.

Tomonaga was born in Tokyo in 1906. He went to the same university Yukawa did, Kyoto University, and actually worked under Yukawa for a while. His interest in quantum electrodynamics was sparked when he visited Heisenberg in Europe in 1937. He spent two years there working under him.

The final episode of the story came in 1949 when Freeman Dyson showed that the formulations of the three men were entirely equivalent. Tomonaga's and Schwinger's equations gave the same graphical rules that Feynman had used in setting up his diagrams. Dyson also completed the proof of the renormalizability of quantum electrodynamics to all orders of approximation. He was able to do this because he worked closely with both Feynman and Schwinger, and was one of the few who thoroughly understood both methods.

In 1965 Schwinger, Feynman, and Tomonaga shared the Nobel prize for their achievements. Of the methods, Feynman's has proven to be the most popular. Feynman apparently laughed to himself when he first developed his little diagrams, thinking it would be funny if *Physical Review* was filled with them. And it has happened. Literally all theorists working in the area now use these diagrams as a guide in setting up their equations.

There is, unfortunately, a sad note in the drama of renormalization. Almost overlooked was a contribution by the Swiss physicist Ernst Stueckelberg. As early as the mid 1930s Stueckelberg had begun to work on a renormalization program. By 1942 he had developed it to the point where he was sure it was publishable and he sent it to *Physical Review*. But because his mathematical notation was unconventional and one of the referees insisted that the program was incomplete, the paper was rejected. Stueckelberg was stunned by the news but determined to make it acceptable. His health was deteriorating now, however, and he managed to work on it only sporadically between periods in the hospital. In 1945 he finished but his work never appeared in a journal and was published only in the thesis of one of his students. But by then Schwinger, Feynman, and Tomonaga had already presented their results.

Physicists were on a high for several years after the successful renormalization of quantum electrodynamics. No theory had ever had such phenomenal agreement between prediction and experiment. Most theorists were sure that it would only be a short time before all such interactions were completely understood. Quantum electrodynamics was the model; all they had to do was extend the methods. Yukawa had developed a theory of strong interactions similar to quantum electrodynamics in 1935. (In place of the virtual photon he had a virtual pion.) Renormalization methods were soon applied to his theory. There were few problems in applying the method but theorists soon discovered that what they got was useless. In quantum electrodynamics second-order terms were 137 times smaller than first-order ones, third-order terms 137 times smaller than second-order ones, and so on. In the strong interactions it turned out that second-order terms were 15 times larger than first-order ones. This meant that the series of terms in the perturbation expansion (i.e., first-order, second-order, . . .) didn't converge—it got larger and larger.

Theorists also tried to apply the same technique to the weak interactions. And again they hit a stone wall. The problems weren't as serious here; nevertheless they were serious enough that no one knew what to do about them. The major problem was renormalization this time; the technique of redefining mass and charge so as to absorb the infinities didn't work here. The infinities couldn't be renormalized away.

The euphoria over the success of quantum electrodynamics had ended by the mid 1950s and field theory had gone into a slump. There seemed to be no way of dealing with the weak and strong interactions. Theorists began to abandon the field theory approach and look for alternatives. Another approach had its foundations in a method introduced in 1943 by Heisenberg. Heisenberg introduced what is called the S or scattering matrix—the collection of all possible interactions. The technique can be applied to either quantum electrodynamics or strong interactions, and since the problems were with the strong interactions, most of the interest centered around the collection of all possible hadron interactions. In S matrix theory we are interested only in the initial and final states; we can represent this simply as

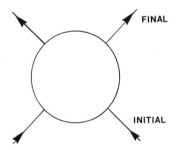

A simple representation of the S-matrix concept.

Theorists worked on the various properties of the S matrix hoping they would show the way. Consideration of the S matrix led to other approaches. One of them was Regge theory, named after T. Regge of Italy. Regge's work was developed by G. Chew and S. Frautschi of the University of California. They discovered that if they plotted the spin of the hadrons against their mass they got straight lines (called Regge trajectories). On the basis of this a model was developed in which the hadrons were all made up of other hadrons. In short, there were no fundamental particles.

Needless to say there was considerable disenchantment at this time and for over ten years little progress was made. In the meantime, however, progress was made in classifying the particles, and in understanding the classification. We will turn to this in the next chapter.

CHAPTER 6

Building a Universe

"QUARK, QUARK"

When Gell-Mann published his eightfold method he was not entirely satisfied with it. True, it was extremely helpful in grouping particles into families and it led to the prediction of the omega-minus particle. But it did not answer the question that was nagging most elementary particle theorists: Are all the recently discovered particles really elementary? Furthermore, Gell-Mann had skipped over the fundamental representation in setting up his theory—and he felt uneasy about it. After all, this was the representation from which all the others were derived; it had to have significance, yet it didn't seem to relate to any of the known particles of nature.

The most logical explanation of the fundamental representation was that it corresponded to a basic triplet of particles from which all others are formed. Sakata had already introduced such a triplet: the proton, neutron, and a particle he called lambda. But they didn't fit into Gell-Mann's scheme. It was easy enough, though, to figure out what the quantum numbers of the particles in his scheme should be. And this was the stumbling block. It was an odd triplet—to say the least. If all other particles were to be made of them they would have to have nonintegral electronic charge. In short, their charges would have to be $1/3$ or $2/3$ that of the electron. And no such charged particles had ever been seen in nature.

Gell-Mann was at an impasse. What should he do? The answer came in March 1963 when he was on a visit to Columbia University. While having lunch with Robert Serber of Columbia the discussion turned to his eightfold method. Why hadn't he made use of the fundamental triplet? asked Serber. It was, after all, the most important representation. Gell-Mann blushingly explained that it wasn't an oversight; he had made the calculations but they didn't seem to make any sense. And he therefore didn't have the gall to publish them. To begin with, most journals probably wouldn't have accepted the paper, and even if one did everyone would think he had gone off the deep end.

Serber listened to his ideas (without laughing) and encouraged him to pursue them. Gell-Mann therefore took another serious look at the nonintegral charges. "What the heck," he said to himself. "It might be possible." They hadn't been observed, but there was a way around this: he would assume that they were trapped inside the particle. The proton, for example, might have three of these particles roaming around in it, interacting with one another but never able to free themselves. He mentioned the idea to Victor Weisskopf upon receiving a call from him. "Oh, come on Murray," said Weisskopf, who was in Europe at the time, "this is a transatlantic call; it's costing money. Let's not pursue this kind of foolishness." And this is more or less the reaction he got from most physicists when he published his paper. He was sure that *Physical Review* wouldn't accept it, but he knew the editor of *Physics Letters*, a journal published by CERN. And, luckily, they were hard up at the time for papers by Americans and eagerly accepted it.

The title of his paper was "A Systematic Model of Baryons and Mesons." And one look at it makes you wonder how such a short paper could be so important. It is barely two pages long and has few formulas in it. Furthermore, as you read through it you can sense Gell-Mann's hesitancy about his strange offspring. He even begins by describing another much less elegant scheme in which the charges are not fractional. Then he introduces them—the "quarks"—with a brief reference to *Finnegans Wake*. He was convinced at this time that any name he gave them would be soon forgotten, so why not add a pun. But to his surprise the name stuck; the joke was on him. He got the name from the line "Three quarks for muster Mark!" in *Finnegans Wake*. To this day we still do not know what the line means; in fact, we don't even know what the word "quark" means. In German it means nonsense, and maybe that's appropriate. Some people have implied that it is the "quark, quark" of a gull. Anyway, whatever it is, it looks like it's here to stay.

One of the lines in the paper caused some confusion. Near the end Gell-Mann said, "It is fun to speculate about the way quarks would behave if they were physical particles of finite mass (instead of the purely mathematical entities they would be in the limit of infinite mass)." Many took this to mean he didn't really believe in them as something real. But as he later stated, "What I meant here was that they would be permanently confined inside the hadrons."

His hesitancy in publishing is also evident in a line near the end of it in which he states, "A search for stable quarks of charge $-1/3$ or $+2/3$ . . . would help to reassure us of the nonexistence of real quarks." If he was sure they were permanently confined it might seem strange that he would encourage experimentalists to look for them. I suppose, though, at this point he's just trying to console anyone who might have thought he had gone bananas.

Well, anyway, now that we know what they are called, what exactly are these quarks? First of all they are, of course, the basic building blocks of the hadrons. According to Gell-Mann, there were three of them, which he referred to as up (u), down (d), and strange (s). (There are also antiquarks in the scheme.) And like all other particles they have quantum numbers; the table below summarizes them.

	J	Q	S	B	I	I_z
u	1/2	2/3	0	1/3	1/2	1/2
d	1/2	−1/3	0	1/3	1/2	−1/2
s	1/2	−1/3	−1	1/3	0	0

According to Gell-Mann all baryons were made up of three quarks, and all mesons of a quark and an antiquark. Of particular importance are the quantum numbers of the quarks: they had to add up to those of the particle itself. And they did.

On paper, then, quarks appeared sound. But in the real world there was a serious problem: they had never been seen. Of course Gell-Mann had hypothesized that they were trapped inside particles. Still, most scientists were skeptical of the whole thing. Assuming they did exist, though, where would you expect to find evidence of them? One place is certainly obvious: among the tracks in bubble chambers. Particle tracks are formed when bubbles condense on the ions that are created as they travel through the chamber. The greater the charge of the particle, the greater the number of ions it produces and the denser or heavier the track. Because quarks have fractional charges their tracks would be easy to distinguish from those created by particles with integral charges.

The photographs of tracks in many different bubble chambers were searched; cosmic rays were also searched. And, perhaps not unexpectedly, they were "found." I say "not unexpectedly" because it seems that whenever something of this nature is searched for it is inevitably "discovered" by somebody. In this case it was an Australian team, and for a while everyone was excited. But as in most such cases, it proved to be a false alarm. When other checked out the experiment they found nothing.

It appears, then, that the most likely places have been of no help. Where else might we expect to find them? What about the earth itself? It is possible that a few stray ones became trapped in rocks during the formation of the earth. With this in mind several geological searches were made. Even seawater, rocks and shells deep in the ocean were examined. But again—nothing.

Actually, there's a more direct way of looking for quarks, or perhaps I should say, a better way of searching for fractional charges. It was used for the first time in 1910 by R. Millikan to determine the charge of the electron, and is now known as the Millikan oil drop method. In the experiment oil is sprayed into an electric field; tiny drops are formed and if the field is properly adjusted some of them remain suspended in it. They remain suspended because each drop has a charge on it—some have one electronic charge, others have two, and so on—and the weight of the drop is balanced by the upward electrical force on the charge. With the weight of the drop and the force required to keep it stationary known, it is easy to calculate the charge it contains.

A variation of this experiment was performed in 1977 by William Fairbank and his group at Stanford University. They used niobium pellets instead of oil drops, but the principle is the same. And indeed they found (or at least believed they found) a charge of $-1/3$. The scientific community was skeptical, but Fairbank had a record of careful experimentation so for a while there was considerable interest. But again the experiment was not verified.

Gell-Mann was asked recently if he believed that Fairbank had actually detected a quark. "If he's found one it's mine," he replied with a chuckle. He has, in fact, said on several occasions he is convinced that confinement is absolute. "But I suppose you can't rule out the possibility of some leakage we don't understand today," he said in a recent interview. He went on to say that if they did find quarks a large quark industry would no doubt grow up quickly around them. The applications would be numerous.

Gell-Mann, it turned out, was not the only one working on quarks. The lack of an explanation of the fundamental triplet had also attracted others. Among them were Yuval Ne'eman and Julian Schwinger. Both made attempts to explain the triplet in terms of fundamental particles but neither succeeded. A third person, however, did succeed. In fact, he came up with the same theory Gell-Mann did—about the same time. To his misfortune, however, he did not manage to get his ideas into print, and most of the glory of predicting quarks therefore usually goes to Gell-Mann. His name was George Zweig.

Born in Moscow in 1937, Zweig came to the United States

while quite young, receiving his bachelor's degree from the University of Michigan in 1959. Upon graduation he went to Caltech to work on a Ph.D., and for three years he struggled with an experimental project that eventually fizzled. Disgusted, he switched into theoretical physics and did a thesis under Richard Feynman. It was during this time that he began reading about the Sakata model and Gell-Mann's eightfold method. And soon he, like Gell-Mann, realized that the fundamental triplet could be represented by a triplet of particles; he called his particles "aces." In 1963 he went to CERN on a one-year fellowship where he wrote up his ideas for publication. Because he was working at CERN he was expected to publish in one of the CERN journal, and as required he submitted his rather long (24 pages) paper to the editor of *Nuclear Physics*.

Zweig was pleased with the way the theory worked out, referring to it as "miraculous." Unlike Gell-Mann, though, he was convinced the fractional charges could be found in isolation and strongly advocated that experimentalists search for them. But nobody took him very seriously, including the editor of *Nuclear Physics*, who refused to publish his paper unless it was severely revised. Zweig worked for a while trying to satisfy him but eventually gave up in disgust. The paper was therefore never published. Despite his misfortune he does have a justifiable claim to codiscovery in that most of his colleagues at CERN were familiar with his work. In fact, news of it even reached beyond CERN: his application for a position at a major university was turned down because the department head thought the model was the work of a "charlatan."

The quark model caused quite a stir, but it was anything but an instant success. Scientists were interested, but skeptical. Fractional charges and confinement were a little too much for most of them. But oddly enough the theory had a lot going for it—it explained many things that could not be explained in any other way. Quarks made sense out of strangeness and isotopic spin, concepts that some thought had been rather arbitrarily introduced. Strange particles were strange because they contained a strange quark. And as for isotopic spin, well, as we saw earlier it can be thought of as associated with an abstract space; if the isotopic spin is in one direction in this space, say the "up" direction, the nucleon is a proton, and if it is in the other direction it is a neutron. In the quark model the proton had two u quarks and therefore had its isotopic spin mainly in the "up" direction; the neutron, on the other hand, had two d quarks, and its isotopic spin was mainly in the "down" direction.

Also, quark theory could explain the long lifetime of the strange particles. According to the theory, strong interactions couldn't change a quark's "flavor" (type); they could only change their spin direction. Weak interactions, on the other hand, could change the "flavor" of a quark. In short, they could change an up quark into a down one. If we compare these two cases it is easy to see why the weak interactions are much slower than the strong ones. In the case of the weak ones a change in quark type is required; in the strong interactions only a spin flip is needed. It is reasonable to assume that much less time is needed to flip a quark than is needed to change its type.

Resonances could also be explained by the quark model. According to the theory quarks could orbit one another, just as electrons orbit the nucleus of the atom. Because there is such a large number of possible arrangements, the number of possible resonances is almost limitless. For three quarks we can have one in orbit around the other two, or two in orbit around one of them. And as more energy is added to the system those in orbit can jump to orbits farther out. The resonance spectrum has, in fact, been completely explained as a result of the model.

Another feature of the theory that was compelling was the symmetry: three quarks and three leptons (the electron, muon, and neutrino) as the only truly elementary particles of the universe. Most felt this was an important feature, as nature seems to prefer symmetry. But then came a shock: a fourth lepton was discovered. Scientists found, much to their amazement, that the neutrinos associated with the electron and the muon were not the same. The symmetry was gone. And Sheldon Glashow of

CHAPTER 6

Sheldon Glashow.

Harvard didn't like it. He was convinced that there should be symmetry and suggested on the basis of it that there was another quark. He even named it, calling it "charm."

Glashow grew up in New York, where he developed an early interest in science, particularly chemistry. As a teenager he built a chemistry lab in his basement where—to his parents' dismay—he performed many dangerous experiments. His father, in fact, was not particularly happy about his interest in science; he wanted him to become a doctor. But he didn't discourage him, thinking it was just a passing fancy that would soon die. But he was mistaken—it didn't die. In fact, it remained high all through his years at Bronx High School of Science during the late 1940s. The science fiction club he belonged to was no doubt responsible for part of his enthusiasm; he even published a few articles on the latest advances in physics in its newsletter. Interestingly, one of his classmates was another Nobel laureate, Steven Weinberg. Yet, strangely, the physics that they were taught while at Bronx High was of questionable value and did little to inspire him. Most of what Glashow learned, he later said, was from discussions with Weinberg and other classmates, and from the numerous books on popular science that he read.

From the Bronx, Glashow went to Cornell. The physics was much better here but the classes didn't overwhelm him with enthusiasm. Most seemed rather dull; besides, he didn't like the way the problems were graded. But then, in his senior year, he took a graduate course in quantum field theory from Silvan Schweber. This was more to his liking, but to his surprise he did poorly on the final exam. Schweber thought he was a little too "cocky" for an undergraduate, and decided to teach him a lesson by giving him an exceptionally difficult final. (If the truth be known it didn't help much.) Anyway, from Cornell Glashow went on to graduate school at Harvard in 1954, but again he was unimpressed with his classes. Then he took one from Julian Schwinger which delighted him; he decided then and there that he wanted to work under Schwinger. But when he went to talk to him about it he found there was a problem: eleven other students also wanted to work under him, and Schwinger wasn't interested in having such a large group. Reluctantly, though, he took Glashow on and gave him a thesis project on the properties of the virtual particles that were responsible for the weak interactions. The work he did on it proved to be of considerable value to him later on.

With the completion of his Ph.D. in 1958 he went to the Bohr Institute in Copenhagen. From there he spent time at Caltech, Stanford, and the University of California, ending up finally back at Harvard in 1966.

Glashow's fascination with charm began while he was at the Bohr Institute in 1964. He was sure a fourth quark that exhibited charm existed. While there he met James Bjorken and explained his ideas to him. Bjorken was skeptical and reluctant to get involved in what he thought was a hare-brained scheme; nevertheless he collaborated with Glashow on a paper that was published in the August 1964 issue of *Physics Letters*. It was a radical idea—but nobody got excited about it. In short, it was just politely ignored.

For the next few years little happened in particle physics. No major discoveries were made and it was generally a boring time for most theorists. In retrospect, though, a number of important papers were published—although they weren't recognized as such at the time. One of these was a paper by Yoichiro Nambu, a Japanese physicist who was born in Tokyo in 1921, received his doctorate from the University of Tokyo, and is now in United States at the University of Chicago. Collaborating with Moo-Young Han he put forward the idea that, besides their usual electrical charge, quarks had another charge. He called it charm, but since Glashow had already used this name for something else it was short-lived. (We now refer to his charge as color.) There were, according to Nambu, three types of this charge. Little attention was paid to the proposal at the time but as we will see later it was eventually realized that it was an important breakthrough.

Considering the lack of progress in particle physics during this time a question that naturally comes to mind is: What was happening in field theory? A few years earlier it had played a central role in theoretical physics. What had happened to it? It had, of course, been tremendously successful in explaining electrodynamics, but when it was applied to the strong and weak interactions it proved to be a failure—and so far no one knew how to get around the difficulties. As a result it had, in a sense, died—or at least almost died. S matrix and bootstrap theory were "in," and they were in conflict, not only with the spirit of field theory but also with the idea of quarks. According to bootstrap theory there were no elementary particles such as quarks; they just weren't needed. A few people were, however, still working on SU(3) (minus the quarks). Indeed, for a while there was considerable activity in the area when it was realized that SU(3) could be generalized to account for two directions of spin. This brought in the group SU(6). Theorists worked feverishly on

SU(6) for a year or so, then someone showed that it was in disagreement with relativity and interest waned. Other groups were tried—almost in desperation. But nothing seemed to pull particle physics out of the doldrums. The major problem was the lack of experimental discoveries. Between 1964 and 1968 there were literally none, and this had a serious effect on theoretical physics. Theorists began grasping at straws, and as a result theoretical physics became splintered into many groups, each working on a different approach, with nobody quite sure which was the right one. Fortunately, a few people continued to work in quantum electrodynamics, and this helped keep field theory alive.

A NEW BEGINNING (THE NEW PHYSICS)

The quark model explained much but what was really needed was evidence that quarks actually existed. It might seem that if they were trapped inside particles it would be impossible to get such evidence. Nevertheless, it came. Unexpectedly, perhaps, but still, it came—from the Stanford Linear Accelerator (SLAC) in the late 1960s.

Situated on land adjacent to the Stanford campus, SLAC was, initially, a controversial project. Scientists referred to it as the "Monster"—a two-mile-long tunnel, not far from the San Andreas fault, inside of which was to be placed a giant linear accelerator. As particles accelerated down the long chamber traffic passed overhead on an interstate highway.

The large accelerator was switched on in 1967 but already experiments were lined up. Teams had been devising ideas for years. The first experiments were perhaps the simplest imaginable: the scattering of electrons off protons. Yet, strangely, they were the ones from which the first evidence for the existence of the quark came. Three groups of scientists were involved in the experiments, one from SLAC, one from MIT and one from Caltech.

Aerial view of the linear accelerator at Stanford (SLAC). (Courtesy SLAC.)

Incidentally, when I say "scattering" I'm not implying that a physical "bumping" occurs, as in the collision of two billiard balls; the interaction is the result of the exchange of a photon. The electrons pass close to the protons and virtual photons pass between them. We can represent it simply as shown in the Feynman diagram on p. 109.

Now to the details of the experiment. They would be of two types, depending on the energy of the electrons. At relatively

View down the tunnel at SLAC. (Courtesy SLAC.)

low energies it was expected that the electrons would scatter *elastically* off the protons; in other words, an electron and a proton would go into the reaction and an electron and a proton would come out. There would be no change in the type of particle involved. This type of experiment was expected to be the

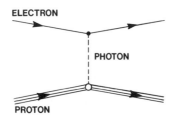

Feynman diagram of a proton–electron collision.

most useful. The overall object was, of course, to determine the internal structure of the proton—if it had any. In particular, they wanted to determine the distribution of charge inside it, although most agreed that it would probably be uniformly distributed.

The second type of scattering is referred to as *inelastic scattering*. It occurs at higher energies. When the energy is sufficiently high the electrons are so powerful that new particles are generated in the collision. The higher the energy, the larger is the number of particles produced. Things can, in fact, get quite messy, and as a result it wasn't expected that this type of experiment would give much useful information. With the large number of particles coming out it would be difficult to determine what exactly was going on inside the proton.

The experiments would, in a sense, be a modern-day version of Rutherford's classic experiment. He too had expected little when he directed Marsden to bombard gold atoms with alpha particles (helium nuclei). But much to his surprise some of the alphas came rebounding back. The SLAC experiments would give a similar surprise.

Scientists weren't quite certain what to expect when they began the experiment, but some were sure that an important breakthrough was possible. It was, after all, the first time they were actually penetrating the proton. To understand what I mean by "penetrate," consider an ordinary microscope. Assume that we want to observe a tiny virus. We know that if we are to resolve it (see it clearly) we must use light with a wavelength much shorter than the size of the virus (if the wavelength is longer we will get nothing but a blur). This is, in fact, why we can't see atoms with ordinary light. Its wavelength is much longer than the diameter of an atom. But when we try shorter-wavelength radiations (e.g., x-rays) we run into a problem: as the wavelength decreases the energy increases. And if we try to look at something small with such an energetic beam we find that it is immediately knocked out of view.

It turns out that we have exactly the same problem with

electrons. As you may remember they also have a wavelength associated with them. And it is again important that this wavelength be much shorter than the objects it is "seeing." In this case it is seeing the proton, and we know the proton has a diameter of about 10^{-13} centimeter. A simple calculation tells us, then, that the beam must have an energy of a few GeV (billion electron volts), and of course the SLAC beam was this powerful.

In the experiment electrons were accelerated down the two-mile tunnel (at close to the velocity of light) and directed at a target of protons in the form of liquid hydrogen. Instruments were set up around the target to measure the number of electrons scattered out of the beam by the protons. The measurements that were of most interest were the energy lost by the accelerating electrons, and the angle at which they were scattered. Also of interest was how much momentum was transferred to the protons. The way electrons scattered from other electrons was well known at the time. If an electron comes close to a direct hit on another electron it is scattered through a large angle. In other words, it rebounds as if it is hitting a "hard" object, in the way a billiard ball rebounds when it hits another billiard ball. But protons are different: they aren't hard objects; they have dimensions. The scattering would therefore depend on how the charge of the proton was distributed throughout it. If it was evenly distributed, as most thought, the scattering would be "soft." In other words most of the electrons would be only slightly scattered, even if they came relatively close to the proton.

The first experiments were low-energy elastic scattering ones. Calculations had been made, so everyone knew what the scattering should look like and indeed everything went as expected. The majority of the electrons were scattered only slightly as they passed through the liquid hydrogen. Granted, a few were scattered through relatively large angles, but this was expected. Everything went so well, in fact, that one of the three teams involved, the Caltech team, decided that little would be accomplished in the remaining experiment, so they withdrew.

The second phase of the experiment began in late 1967. Things were expected to get messy now: new, heavy particles would be coming out in the inelastic collisions that would be occurring. Such a mess was a theorist's nightmare. With little to guide them it seemed it would be virtually impossible to make calculations that made sense. It was expected that resonances or excited states of the proton would be detected at lower energies, but beyond that at higher energies things were uncertain. Most felt sure, though, that there would be few electrons scattered at large angles.

The experiment took place and the following spring the data were analyzed. At low energies (within the inelastic region) everything went as expected: the resonances were clearly visible on the plots as bumps. But when they began looking at higher energies they got a surprise: the resonance peaks were gone (as expected), but the cross-section curve did not decay off as expected. It remained high. (This section of the plot is now referred to as the "continuum.") What did this mean? There seemed to be only one answer: the distribution of charge inside the proton was not smooth, but rather in the form of "hard" point charges. Most experimenters were not ready at this stage, though, to accept this interpretation. There were other possibilities. The electrons would radiate a lot of their energy away upon being deflected. Maybe this was the problem. But when the appropriate calulations were made they found to their dismay that radiation losses couldn't explain the effect.

One of those involved in the calculations was James Bjorken, who had joined SLAC at its initiation in 1963, before the accelerator itself was built. While it was being built he spent his time trying to predict the outcome of the first experiments. Now, with the data coming from the experiments he was in the thick of things. He began making plots of various types and soon discovered something strange. When the momentum or energy of the electrons became large the dependence of the cross-section on the momentum and energy became very simple. The name that became attached to his discovery was "scal-

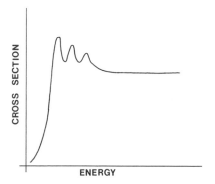

A plot of cross section with increasing energy.

ing." Nobody was quite sure what this scaling meant; even Bjorken was unsure, but he did come up with an explanation using complicated mathematics. The problem was that nobody understood what he was doing.

Then came Richard Feynman. Feynman had continued working in the area of elementary particles for a few years after he made his monumental discoveries in the late 1940s, but he had eventually drifted into general relativity. One of the major problems of the day (and also of today) was quantizing Einstein's theory, and Feynman was interested in trying his hand at it. Like many others before him, however, he soon saw the magnitude of the problem, and after a few years decided it was time to get back into particle physics. He asked some of his colleagues about the current problems and was told about the SLAC experiments. He decided, therefore, to visit SLAC to see what was going on. At this stage he had never met Bjorken, and had never heard of scaling.

The laboratory was in a flurry of excitement when Feynman arrived. So much new data—but nobody could explain them. Feynman decided to chuck his hat into the ring. He began by trying to visualize what was going on in the experiment. Let's see if we can do it. We'll begin by assuming that the proton and

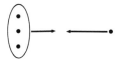

How the electron "sees" the proton (i.e., as a pancake). Feynman assumed the electron would interact only with one of the partons, and not with the proton as a whole.

electron are flying toward one another at close to the speed of light. In the actual experiment the proton is sitting still relative to us, but according to Einstein all motion is relative so we can equally well think of it in this way. The electron is a point particle so its velocity has no effect on its dimensions, but the proton is finite, therefore, as seen by the electron during the passage it will look like a pancake—flattened because of the effects of relativity. But relativity also tells us that time will be slowed down; therefore, anything inside it will be frozen (appear motionless).

What, indeed, would we expect to be inside it? Feynman wasn't sure, but he knew from quantum electrodynamics that electrons emit photons, which in turn give electron–positron pairs. Each of the resulting particles could, in turn, give further photons and pairs. In short, an electron was really an electron cloud. In the same way protons emit virtual pions that produce pairs, and each of these pairs is capable of giving pions and pairs. Thus, the proton could also be thought of as a cloud of particles. There was, of course, the possibility that there were quarks involved. After all, Gell-Mann had given convincing evidence that there were quarks inside protons. But Feynman ignored them; he decided to call the particles partons. He didn't care exactly what they were—that wasn't important to his analysis.

In short, then, the electron would "see" a pancake-shaped proton full of particles—partons—that would be frozen in place (see figure). But it was Feynman's next step that was the critical one. He assumed that the virtual photons emitted by the passing electron would interact with only one of the partons. This was a bold move, for it allowed him to consider a photon–parton interaction rather than a photon–proton one—a situation that was much easier to deal with.

Much to Feynman's delight he found that he could explain Bjorken's scaling. Bjorken had partially explained it in a complicated way but now we had a simple explanation. And even Bjorken accepted it. Shortly after Feynman left, in fact, he published a paper on partons and scaling with a colleague. Feynman had not yet published his model at this stage (that didn't come until 1972) but Bjorken credited him with it in his paper.

Everyone was pleased with the new parton model. Everyone, that is, except Gell-Mann. "Why partons?" he asked, in annoyance. He had shown that there should be quarks inside protons. Where did these partons come from? Interestingly, Feynman's office at Caltech is right next to Gell-Mann's, and with Gell-Mann's disgust with partons, relations between the two men were cool for a while, to say the least. In the back of most people's minds, however, partons were associated with quarks. Yet according to Feynman's theory, they didn't seem to be quarks; there were several problems. One of them was that the partons were assumed to be free; in other words they didn't interact with one another in Feynman's theory. But quarks did interact with one another. It also seemed that a mere three quarks was too few to explain scaling via the parton model. To get around this, two classes of quarks were assumed to exist. Just as we visualize the proton to be surrounded by a cloud of pions and nucleons, so too could we visualize it (in the quark model) to be surrounded by a "sea" of quarks. In short, the two classes were the three usual quarks, and the surrounding "sea."

Calculations were again made under this assumption and there were still problems. But of course no one had yet taken into consideration the interaction between the quarks. How did they interact? The idea that there was an exchange particle (later called a gluon) had been around for several years. In 1971 Victor Weisskopf picked up on the idea and developed a detailed theory. In his model virtual gluons were exchanged between quarks, just as virtual photons are passed back and forth in electromagnetic interactions. This brought the theory into accord with experiment, and, indeed, today everyone accepts the fact that partons are quarks.

More proof of the model was needed, though, and it soon came when the electrons were replaced by neutrinos. The neutrino–proton experiments were much more difficult to carry out than the electron–proton ones because of the elusiveness of the neutrino. It rarely interacts. But with patience and perseverence it was shown that the scaling phenomenon also occurred here and it could be explained by the parton–quark model.

The existence of quarks had at last been verified. Pieces of the puzzle were finally beginning to come together. The hadrons of the world were made up of quarks; the quarks were the truly elementary particles.

But for a complete theory of the universe we also have to explain the fields that hold particles together and in the last chapter we saw that by the mid 1950s field theory was in trouble. Let's turn back now to field theory to see how the problems were overcome.

CHAPTER 7

Gauging the Universe

Breakthroughs in science sometimes result from things so commonplace it's almost an embarrassment to scientists when the discovery is finally made. Such was the case in field theory. The commonplace phenomenon was symmetry. The recognition after many years that symmetry plays an important role in nature was the key that opened the door to some of the most important discoveries that have been made in the last couple of decades.

Symmetries are, of course, all around us. You can easily recognize the symmetry of a leaf: its left side is the same as its right side. Similarly, the star shown below has symmetry. If you rotate it about its center so that point A goes to point B the figure stays the same. This is, in essence, what symmetry is all about. An object is symmetric if it doesn't change when rotated or transformed in some way. We have rotated the star through an angle of 72 degrees and it has remained the same. We say it is "invariant" under rotation.

Looking around, you see many other symmetries. Your body, for example, is symmetric. A square box is symmetric, as is a baseball. These symmetries are called geometric symmetries. They are an important type but they aren't the only type. There are other cases where, after a transformation is made, the system remains the same. Take a magnet, for example. You are likely familiar with the fact that it has a north and a south pole and that lines of magnetic force run between the two poles. But if we interchange the poles there is no change in the form of the

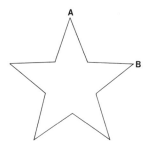

A star—a symmetric object.

magnetic field lines; they stay the same. The same thing applies to the lines of electric force that run between a positive and a negative charge.

Another less familiar symmetry, one that has been invaluable to theorists in recent years, is one associated with isospin. Isospin, you may remember, was introduced to account for the difference between the proton and the neutron. In an imaginary isospin space we have only one particle, the nucleon, which is a proton or neutron, depending on how an arrow is pointed in the space. The symmetry in this case is what is called continuous symmetry—similar to the kind a ball has. (Regardless of which way you turn the ball it always looks the same.)

Symmetry is important because it is connected with the basic laws of physics—the conservation laws. The law of conservation of energy, for example, says that the energy is the same after a reaction as it was before. But the real breakthrough in understanding the importance of symmetry came when the German mathematician Emmy Noether pointed out that there is a symmetry associated with each of the conservation laws.

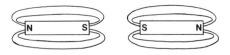

Magnetic field lines around magnets.

A stocky woman who wore thick glasses and cared little about how she looked, Noether led a tragic and rather short life. Born in Germany in 1882, her ability in mathematics was evident at an early age. And with the help of her father, a mathematics professor, she obtained a Ph.D. in 1908. This, in itself, was an accomplishment in that very few women were allowed to attend a university at that time. But even with the degree her life wasn't easy. She was unable to get a paying job. Her love of mathematics was so strong, though, that she offered her services free to a nearby mathematics institute. Her first official teaching position was at Göttingen—but again it was an unpaid position. Not until the early 1920s, some ten years after she had acquired her Ph.D., was she finally paid.

Noether was not highly successful as a teacher, mostly because of her unorthodox approach and appearance: she spoke with a loud voice, never combed her hair, frequently waved her hands wildly, and never backed down from an argument. To the student of the day she was no doubt something of a phenomenon. But her insight into complicated mathematics was phenomenal. She had both an uncanny ability to penetrate to the heart of a mathematical problem, and an intense desire to understand fundamental problems.

Noether's stay at Göttingen was, unfortunately, short. Naziism was on the rise in Germany and within a year or so she and several others had to flee for their lives. She emigrated to the United States and after another long struggle managed only to get a temporary position at Bryn Mawr University in Pennsylvania. Two years later she died at the age of 53.

Her theorem stating that symmetries and conservation laws are connected attracted little attention for many years. Mathematicians were interested only in her contributions to group theory, number theory, and other abstract branches of mathematics, and most physicists had never heard of her.

To see a simple application of her theorem consider space. Yes, plain, ordinary space. No one would argue that it is not symmetric; it's the same everywhere, and the laws of physics are therefore the same everywhere within it. This means there is a law of conservation associated with it. Which one? It is easy to show that it is the conservation of momentum, in other words the law that says that momentum (mass times velocity) is the same before a reaction as it is after.

Symmetry has, in fact, now become so important in physics that whenever we find a new symmetry in nature we immediately ask with what conservation law it is associated. Or conversely, if we discover a conservation law, we try to find the associated symmetry. Consider the conservation of charge—the law that says that the electrical charge in any reaction has to be the same before and after. With what symmetry is it associated? Answer: gauge symmetry—a type with which you are not likely familiar. It was introduced by the German mathematician Hermann Weyl shortly after Einstein published his general theory of relativity in 1915. Weyl used it in his attempt to unify Einstein's theory of the gravitational field and Maxwell's theory of the electromagnetic field.

What do we mean by a gauge? To gauge something is, of course, nothing more than to measure it. To set a gauge therefore means to specify the measuring rod. Once it is specified, we assume, naturally, that its length will remain constant. It would seem strange if its length varied as we moved it from point to point, and in fact if it did the whole idea of measurement would make little sense. Nevertheless, Weyl considered this possibility—he referred to it as a local gauge change. Changes that were the same everywhere were referred to as global gauge changes. A simple example of a global change would be to move every house in a town fifty feet to the right. No change would be evident between one house and another; the movement would be the same for each house and would leave the town unchanged. If, however, we were to move the houses at random, some to the right, some to the left, we would have an entirely different situation—in short, we would have chaos. This is a local change, and at first glance it would seem that such a change would be of no importance in physics. Not so. As odd as it might seem, Maxwell's theory of electromagnetism has this property. To understand how this occurs, consider a number of electrical charges, each of different electrical potential (voltage); some of them positive, others negative, and assume they are moving around at random. If we add, say, fifty volts to each of them we know there will be no overall change; the difference between any two will be the same (i.e., $60 - 55 = 5$ and $10 - 5 = 5$). This is a global change. If we were to add a different voltage to each, local changes would occur. This means that the electric field associated with the charges does not satisfy local gauge invariance. But if we look closer, we see that something phenomenal happens: the moving electrical charges create magnetic fields, and associated with these fields are what are called magnetic "potentials" that exactly compensate for the change. The result: the combined electric and magnetic field—the electromagnetic field—*is* gauge invariant.

Weyl was, of course, familiar with this and hoped that something similar would allow him to unify, or join, the gravitational and electromagnetic fields. He modified general relativity with this in mind and, as he had hoped, Maxwell's electromagnetic theory appeared. It was like a miracle and Weyl was certain he had achieved a unification of the two fields. He immediately rushed a copy of his paper off to Einstein. But it didn't take Einstein long to pinpoint a flaw. He showed that if the idea was valid, clocks moving relative to one another would run at different rates. If, for example, we had two observers, each with a clock (we assume the two clocks run at exactly the same rate when they are together), and if one of them moved off, taking a trip to a nearby town and back, the elapsed time on the two clocks would not be the same when the observers got back together. Weyl was disappointed when he heard from Einstein, but realized he was right. His theory was, indeed, flawed, so he laid it to rest.

But good ideas sometimes die hard. And this was a good idea. It had just been applied to the wrong theory. de Broglie

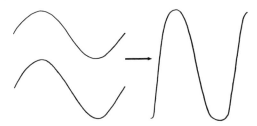

Waves shown in phase with one another. In this case they reinforce one another. If they are out of phase they cancel one another.

had shown that electrons have a wave associated with them, and Schrödinger had incorporated this wave into his quantum theory. As we saw earlier, he introduced a "wave function psi." One of the important properties of any wave is what we call phase. We know, for example, that when two light waves are in phase (i.e., when their "humps" are in unison) they reinforce one another. If they are exactly out of phase, on the other hand, they cancel one another and we get no light. According to Schrödinger's theory electron waves had a phase, yet strangely in the final analysis this phase didn't affect the predictions of the theory. It was shown that this occurred because the local gauge symmetry of the theory compensated by creating photons. The quantum theory in which we have both electrons and photons interacting is, of course, quantum electrodynamics (QED). So we can say that QED, like electromagnetic theory, also has local gauge invariance.

But if local gauge invariance is so important in these theories, what about the other fields of nature, say the strong interactions? Are they locally gauge invariant? In the early 1950s Chen Ning Yang of the University of Chicago decided to see if he could set up a theory of the strong interactions that had local gauge invariance. He felt that such a theory might lead to a better understanding of these interactions.

Yang, who could still pass for a graduate student when he

Chen Ning Yang.

received his Nobel prize, was born in Hofei, China in 1922. When he was 5 his family moved to Peking because of the war with Japan, but the war soon caught up with them and they had to move again. He studied at the National Southwest Associated University, obtaining his masters degree in 1944. In 1945 he came to the United States to work under Fermi at the University of Chicago, completing his Ph.D. there in 1948. While at the University of Chicago he and another Chinese physicist, Tsung Lee (who later shared the Nobel prize with Yang), had a class taught by the well-known astronomer S. Chandrasekhar. They were, in fact, the only two students in the class. Chandrasekhar referred to the class later "as the one in which everyone in it won the Nobel prize."

Yang's initial attempt to make the strong interactions into a local gauge theory was unsuccessful, but he kept the problem in mind when upon graduation from the University of Chicago he went to the Institute of Advanced Study at Princeton. There he again tackled the problem but was still unable to solve it. The major difficulty centered around the type of symmetry he would have to impose upon the system. In 1953 he took a leave of absence and went to Brookhaven National Laboratory. Most of his time at Brookhaven was taken up predicting the outcome of the experiments that were being performed. But the gauge problem kept gnawing at him. He mentioned it to his office mate, Robert Mills, and together they tackled it.

They decided the best symmetry to use would be isospin. With this as a starting point they modeled their theory on quantum electrodynamics, introduced local gauge invariance, and, as expected, out popped a new force field. They referred to it as the B field; it was analogous to the photon field of quantum electrodynamics. And of course they hoped this field would have the same properties as the strong nuclear field. But it didn't. The strong nuclear field was extremely short-ranged (it acted only over a distance about equal to the size of the proton) and this meant that the particle representing it had to be very massive. The gauge particles (they are also called exchange particles) that appeared in the Yang–Mills theory, on the other hand, had no mass. Furthermore, not one appeared, but three.

Yang and Mills worked to make sense out of the theory; in particular they wanted to make physically relevant predictions using Feynman diagrams. But the more they tried the worse things got. They finally gave up in frustration, but felt the idea was interesting enough to publish. And in 1954 their paper appeared in *Physical Review*. Despite its shortcomings it is now considered to be a classic.

The Yang–Mills gauge, or exchange, particles were strange in another respect. We know that the gauge particle of the electromagnetic field, the photon, does not interact with other photons. The Yang–Mills particles, on the other hand, did interact with one another, and this meant that they had Feynman diagrams involving only gauge particles.

Scientist were still excited about field theory in 1954 (the doldrums of the 1960s had not yet come). So there was considerable interest when the Yang–Mills theory appeared, and many theorists followed up on it. The major problem was, of course, the massless gauge particle. But there was a way around this, an unsatisfying way—but a way nevertheless. You could just add in a mass "by hand"—in short, just put a mass term in the equations. And this is what a number of theorists did. The problem with this—and it was a serious problem—was that it destroyed the gauge invariance of the theory and made it unrenormalizable.

Eventually interest in the theory waned and for many years few gave it a second glance. It was a novelty but nothing more. Fortunately, this isn't the end of the story. To continue, though, we must first look at the weak interactions, since it was in these interactions that the above problem was first overcome and a satisfactory gauge theory was formulated.

THE WEAK INTERACTIONS

The discovery of what we now know as beta decay was made by Antoine Becquerel in 1896. It marked the beginning of the study of the weak interactions but, of course, it was many years before scientists realized this.

Ernst Rutherford saw the strange rays even before Becquerel, calling them beta rays. At the same time he noticed another type he called alpha rays. He later showed that the alpha rays were helium nuclei. But it was Becquerel who showed us that beta rays were electrons.

Becquerel was trained as an engineer. His father, a professor of physics at the Paris Museum of Natural History, did important early work on fluorescence. Antoine eventually succeeded him. The turn of the century was an exciting time for physicists: Wilhelm Roentgen had galvanized the European scientific community with his discovery of x-rays. Everyone, including Becquerel, was fascinated with the new phenomenon. Becquerel's most important discovery occurred, as so many discoveries do, almost by accident. He decided one day to see if the rays from certain fluorescent chemicals would penetrate black paper. He began by placing the material in sunlight (presumably to make it fluoresce, or glow) with a sheet of black paper and photographic paper under it. When he developed the photographic paper he found it was fogged: the rays had penetrated the black paper. He was then interrupted by a series of cloudy days, so he put the fluorescent material (wrapped in black paper and photographic paper) away in a drawer and waited for the sun. But the clouds continued . . . and continued. He finally got so exasperated he developed the photographic paper anyway, just for the heck of it. To his amazement it was heavily fogged. What was going on? It was fogged even without the sun! Sunlight must have nothing to do with the radiation. Determined to find out what was happening, he soon discovered that some of the rays that were fogging the paper were like Roentgen's x-rays—but some were not. He put the source in a magnetic field and found the rays were bent in a direction indicating that they were negatively charged. In 1900 he finally concluded that they had properties identical to those of J. J. Thomson's cathode rays. In other words they were electrons.

What was happening was that the neutrons in the material were decaying into protons and electrons, a process we now refer to as beta decay. At the time scientists were unsure where the electron and proton were coming from so they never worried when they discovered that the electrons were coming out with a large range of energies. But when they realized that it was the neutrons that were giving rise to them, a problem immediately presented itself: the electrons should all be coming out with the same energy. Why weren't they? The answer came from Wolfgang Pauli: he said that there was another particle being emitted, one we could not see, a particle we now call the neutrino.

By the early 1930s scientists finally began to realize that they

Enrico Fermi.

were dealing with a new force of nature and in 1934 Enrico Fermi, at the University of Rome, formulated the first theory of the weak interactions.

Born in Rome in 1901, Fermi was so sickly a child that he spent the first two years of his life being cared for by nurses in the country. And it is perhaps surprising that he kept his health when he was brought home. His father was a railroader and the family lived in a drafty, unheated apartment near the railroad station. It was cold so much of the time that his older brother caught cold and died of pheumonia when he was 15. Enrico and his brother had been close and the death was a tremendous shock to him. To ease the sorrow he buried himself in books, and this led eventually to a fascination with mathematics and physics. He began spending all his money on secondhand physics books, and the more he read the more enthralled he became. "See this . . . it is beautiful," he would say to his sister, pointing to some mathematical proof. When he graduated from high school he applied for a fellowship at a university for superior students at Pisa. Part of the entrance requirement was a paper on a technical subject. Fermi's paper was on vibrating strings, and it impressed the judges so much they asked him to come in for an interview. When he entered the university in 1918 he was already far ahead of most of his classmates. But this didn't stop him from studying on his own, and soon, to the dismay of his professors, he left them behind. Despite his brilliance, though, he wasn't beyond high jinks and almost got himself expelled for letting off a stink bomb in one of his classes. Part of the problem, no doubt, was that he was bored.

To his surprise one of his professors called him into his office one day. "Will you teach me physics?" the professor asked. "Teach a professor?" said Fermi in surprise. The professor nodded, admitting that he knew far less than Fermi. A class was arranged.

Fermi received his doctorate in 1922; his thesis on x-rays was so well done and so far beyond most of those on his examination committee that they didn't know what kind of questions to ask him and quickly passed him. Upon completion of the degree he went to the University of Rome and within four years was a full professor.

At Rome he performed a classic experiment in which nuclear fission (a splitting of the nucleus) was produced, but was so intent on looking for something else that he missed the discovery. Someone once asked him if it was a great disappointment to have come so close. "No, I'm glad I missed it," he replied, nonchalantly. But of course he didn't miss further developments in the field, and eventually performed the most famous experiment of all: the first sustained fission reaction. This occurred on a dismal day in December in the squash courts beneath the stands at Stagg Field at the University of Chicago. The building of the atomic bomb at Los Alamos soon followed.

Fermi wrote his famous paper on the weak interactions while at the University of Rome. He told several colleagues about the theory one day after skiing in the Alps. "I think this is the paper I will be remembered for," he said with pride. He was

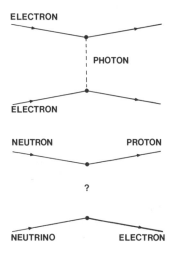

Upper panel shows QED interaction of two electrons. Lower panel shows similar weak interaction.

sure it was the best thing he had ever done. As many others had done before and after him, he modeled the theory on quantum electrodynamics. As you know, when we have two electrons passing close to one another, photons are exchanged. Fermi assumed that in beta decay the neutron changed into a proton at the interaction point, and at the same time the neutrino changed into an electron. The particle that took the place of the photon, in other words the exchange particle, had to be extremely short-ranged so Fermi assumed that the interaction took place at a point. The decay would then be represented as shown in figure on p. 130.

Despite his confidence in the new theory, Fermi's paper describing it was rejected by *Nature*. The editor returned it with a short note saying that it had little to do with present-day physics and would therefore be of no interest to most physicists. Stunned, Fermi sent it to an obscure Italian journal (and later to

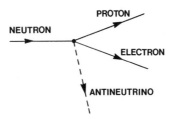

Neutron decaying to a proton, an electron, and an antineutrino (beta decay).

a German journal). It therefore began with a "whimper," but it eventually announced itself loud and clear: it became the accepted theory of the weak interactions. And indeed, with few changes it is the theory we use today. Some have said that it was worth a Nobel prize. Of course he didn't get it for this paper, but he did win it a few years later.

The theory was not, however, problem-free—far from it. It was, in fact, incomplete: it did not explain why the field was weak, and it said very little about the nature of the weak force. Furthermore, it was only valid at low energies.

PARITY

Another flaw in Fermi's theory didn't become apparent for several years. Earlier we talked about symmetry. One type I didn't mention is called reflection symmetry, or parity. It is the symmetry that occurs upon exchange of right and left. If you look in the mirror you know that the image you see has right and left interchanged. Since you generally look the same on the right side as you do on the left you notice little difference. Your image conserves parity. (To be perfectly accurate parity also involves a further reflection, but we'll ignore it for now.)

Parity can also be applied to nuclear reactions. If the mirror image of a reaction occurs we say the reaction conserves parity. And for many years this was assumed to be the case for all interactions of nature. But soon after the strange particles were discovered an oddity involving the kaons was noticed. There appeared to be two types of kaons referred to as tau and theta that had exactly the same mass and spin, but decayed differently. They were so similar, in fact, that a number of scientists wondered if they were, perhaps, the same particle. Yet that was impossible. They could only be the same if the decay process did not conserve parity—and everyone knew that parity was conserved. Or at least they thought they knew. The problem became known as the tau–theta puzzle.

In April, 1956 a meeting was held in Rochester, New York. The tau–theta puzzle was to be on the program. At this stage theorists were still thoroughly confused. Gell-Mann and Yang both gave short talks on the problem, but were unable to make any suggestions for its solution. In the audience was Richard Feynman, and sitting next to him was an experimentalist from Duke University, Martin Block. Feynman began using Block as a bouncing board for some of his ideas, but soon the two were in a heated argument. "Well, maybe parity isn't conserved," said Block suddenly. Feynman started to tell him it was a stupid idea, then stopped. Maybe it wasn't so stupid: perhaps parity wasn't conserved.

Yang and a colleague, Tsung-Dao Lee of Columbia University picked up on the idea. They had met earlier in China when both were students at the National Southwest Associated University. Lee had come to the United States on a scholarship in 1946. Like Yang he studied at the University of Chicago, receiving his Ph.D. in 1950. He joined the faculty of Columbia University in 1953 and soon received several distinctions: At 29 he was the youngest person to ever become a full professor at Columbia. And shortly thereafter he became the first Chinese (along with Yang) to receive the Nobel prize. Furthermore, he was the second youngest to ever receive it. Lee says he "spends most of his time thinking" but admits he occasionally reads mystery novels.

Yang and Lee soon convinced themselves that tau and theta were the same particle, and therefore parity had to be violated in

Tsung-Dao Lee.

the weak interactions. Yet they were sure that someone had proved that it is conserved. Much to their surprise, though, when they made a thorough literature search they found that no one had. It had just been taken for granted.

In 1956 they published a paper in *Physical Review* presenting their evidence. But of course a check was needed—and it soon came. The following year Chien-Shiung Wu, a Chinese physicist at Columbia University, announced that she and Ernst Ambler, of the National Bureau of Standards, had checked, and indeed, parity was not conserved in the weak interactions. Within a year Yang and Lee were awarded the Nobel prize.

Wu had the jump on most other experimenters in that she was an expert in the weak interactions and was in frequent touch with Yang and Lee. It was, in fact, Wu who Yang approached to find out if anyone had proven that parity was conserved by the weak interactions. She didn't know but suggested that he do a literature search.

Even before Yang and Lee had published their paper Wu was hard at work. She had decided to use a radioactive form of cobalt and look at beta decay. What she had to do, in essence, was see if the electrons came out symmetrically with respect to the nuclei, or asymmetrically. If they came out asymmetrically, parity was not conserved. In order to see asymmetry she would have to line up the nuclei (make them all point in the same direction). Each of them had a tiny magnetic field so she was able to do this by placing them in an exterior field. But even with this there was still a problem: at room temperature nuclei move around continuously, and the strongest magnetic fields available would not keep them aligned. She therefore had to cool them to near absolute zero to keep them still. With patience and hard work she was finally able to complete the experiment. She then double-checked to make sure there was no doubt. Then in January 1957 she announced her results: Yang and Lee were right.

Parity (P) was not conserved. But strangely if you combined parity with charge conjugation (C)—a process in which you change the sign of the charge—the combined process CP did appear to be conserved. And for many years sceintists were convinced that CP was conserved. But in 1964 V. L. Fitch and J. W. Cronin of Princeton University decided to check on this, and indeed they found that not even CP was conserved exactly.

A question that now comes to mind is: How did the nonconservation of parity affect Fermi's theory of beta decay? Fermi obviously hadn't taken it into consideration. And indeed it did cause a problem. To understand the problem we must look at the nature of fields. Fields are classified according to the number of variables that are needed to describe them. The simplest field is the scalar (S) field, a good example of which is the temperatures in a given area. They can be specified by one number for each point. For the electromagnetic field it turns out that four numbers are needed, and it is therefore known as a vector (V) field. The gravitational field, on the other hand, is particularly complicated; ten numbers are required for it. It is referred to as a tensor (T) field.

Fermi had modeled his theory on quantum electrodynamics and as such it was also a vector field. But once it was known that the weak field didn't conserve parity, scientists realized it couldn't be a pure vector field. E. C. Sudarshan and Robert Marshak of the University of Rochester eventually showed that it was a combination of a vector field and what we refer to as an axial–vector field (roughly speaking this is a vector field that does not conserve parity). More technically it was V − A (vector minus axial vector). Fermi's theory obviously had to be modified to adjust for this. And the modification soon came. Gell-Mann had, for several years, been interested in developments in weak field theory. Furthermore, the professor in the office next to his at Caltech, Richard Feynman, had also developed an interest in it.

Feynman heard of the Yang–Lee result at the Rochester Conference where Lee had given a paper on it. But he was confused. He read the paper Yang and Lee had written and couldn't make much sense out of it. "I can't understand these things Lee and Yang are saying . . . so complicated," he said to his sister the evening after the conference.

His sister laughed. "No, you don't understand it because you didn't invent it," she said. "Figure it out your own way . . . then you'll understand it." He took her advice and sure enough it finally came clear. He had, in fact, worked on something similar years before.

That summer he went to Brazil. When he returned he went to Wu's laboratory at Columbia to find out what was going on in beta decay but she wasn't there. Someone in her lab tried to explain the latest developments to him but he couldn't understand them so he returned to Caltech. At Caltech three colleagues began giving him details of the theory that had been developed by Marshak and Sudarshan (and independently by Gell-Mann). As they talked it suddenly dawned on him: "Weak theory had to be V − A. Everything fit!" He quickly got together with Gell-Mann and they soon had a consistent revision of Fermi's theory that incorporated V − A. A paper was published in *Physical Review* in 1958.

The theory had been repaired. Yet it was still far from a perfect theory. Feynman and Gell-Mann had introduced parity nonconservation into it, but they still assumed that the interaction took place at a point. There was no exchange particle. Furthermore, the theory was unrenormalized and there were problems at high energies.

Something else was needed. The theory wasn't as successful as quantum electrodynamics because it was still different in an important respect: it was not a gauge theory. Yet scientists refused to believe that gauge was an important aspect of physical theories. It was just an interesting, perhaps accidental, quirk of quantum electrodynamics.

But maybe the difficulty with the weak theory occurred because it was, in reality, part of the electromagnetic theory. Was there, indeed, a way to unify the two theories? They were similar in many respects, but there were significant—very significant—differences. And these differences deterred most physicists from attempting a unification. One they didn't deter was Julian Schwinger. In 1958 he published a paper titled "A Theory of Fundamental Interactions" in which he combined the two fields. He was sure, in fact, that he had properly unified them.

Several years earlier the German physicist Oskar Klein had made the suggestion that there was an exchange particle associated with the weak interactions. He called it W, the same letter we use today. Schwinger incorporated Klein's idea into his theory, assuming that there were two W's, a positive and a negative one. They, along with the photon, were put in a family. He chose the group SU(2) as the basis of his theory, assuming that the electron and the neutrino were like the proton and neutron. In other words they were the same particle and took on a particular identity (electron or neutrino) depending on how an arrow in an isospin-like space was oriented. Beta decay in Schwinger's theory was represented as

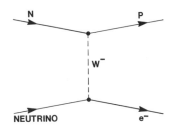

Beta decay in Schwinger's theory.

Schwinger's theory was interesting but it had a fatal flaw. He had not yet heard that theorists had shown that the weak field was V − A; he used a combination of vector and tensor (a tensor is a higher order vector) fields. When he heard the news—after his paper had been published—he was annoyed. So annoyed, in fact, that he did no further work on the theory.

Shortly after Schwinger published his paper a theorist at Lawrence Radiation laboratories, S. A. Bludman, invented a theory that did incorporate the V − A interaction. Furthermore it was a gauge theory. But unlike Schwinger's theory it did not include both the weak and the electromagnetic fields; it was a theory of the weak interactions only. The major difficulty of the theory was again a massless exchange particle. If it was to have a mass it had to be put in "by hand" and this destroyed the gauge invariance of the theory. Nevertheless it was a step in the right direction.

Schwinger swore off the weak interactions but he did pass the torch to his student, Sheldon Glashow. Glashow had begun thinking about the weak interactions while doing his Ph.D. thesis. And from 1958 to 1960 he worked feverishly trying to effect a unification. As an ardent admirer of gauge invariance and renormalization, he believed they were an important part of any theory of interactions. Thus his theory differed from Schwinger's in that it was a gauge theory.

In early 1959 he thought he had finally accomplished his goal: a unification of the electromagnetic and weak fields. With

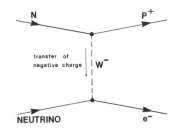

A weak interaction involving charged currents.

pride he rushed to show his ideas to Abdus Salam in England (Salam was also working on the same problem) but to his dismay Salam promptly shot them down. Salam showed that there were several serious mathematical errors in the paper. Glashow was embarrassed, but he refused to give up. In fact, he became even more determined to accomplish his goal. He attacked the theory with renewed vigor and in 1961 published the first of a series of papers that was, indeed, on the right track.

At first, though, it seemed that even Glashow's new theory had a flaw. In addition to the known exchange particles of the weak interactions ($W^{\pm}$) it implied that there was a neutral particle Z^0. This meant that there were "neutral currents" in the weak interactions, and such currents had never been observed.

Let's look at what we mean by neutral currents? Consider a Feynman diagram for beta decay; as you know it involves the W^-. Furthermore, it is easy to see that a charge is transferred from one side of the diagram to the other via the W^-. (Other similar diagrams involve the W^+.) In the case of the Z we would have a diagram such as

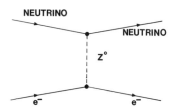

A neutral current intereraction.

In this case there is no transfer of charge. It is referred to as a neutral current interaction, and at the time Glashow published his paper no such interactions had been observed. But this wasn't the only thing theorists didn't like about Glashow's theory. Another unsatisfactory aspect was that he coupled the weak and electromagnetic fields together using a "mixing angle." Furthermore, masses were introduced artificially as in earlier theories. The worst thing about the paper, though, and perhaps what led to it being generally ignored, was that it was written in Schwinger's odd notation. Few people were familiar with it.

Across the Atlantic in England Abdus Salam was also busy working on a similar theory. Born and raised in British India, Salam left for England in 1946. Upon completion of an undergraduate degree in mathematics at Cambridge he decided he wanted to learn some physics. What he was really interested in was theoretical physics, but his marks were so high he was encouraged to go into experimental physics. (Since Rutherford's day it had generally been accepted that the best students would go into experimental physics.) So Salam reluctantly began working on an experimental project dealing with the scattering of tritium and deuterium (heavy forms of hydrogen). He was discouraged to find out how archaic the labs at Cambridge were. Furthermore, there were no technicians to build equipment or blow glass—you had to do everything yourself. This part didn't bother Salam as much as finding out that when he completed a piece of apparatus it never worked. Nothing seemed to work. He grew increasingly frustrated and finally admitted to himself that he was not, and never would be, an experimentalist. With a sigh of relief he switched into theoretical physics and worked on quantum field theory under Nicholas Kemmer.

Upon completion of his doctorate he returned to Pakistan to teach at the University of Lahore. But to his dismay he found that his colleagues had no interest in quantum field theory and particle physics. And research was unheard of. He tried to work on his own, but soon realized that unless he returned to England, he would get nowhere. So back to England he went,

Abdus Salam.

eventually securing a professorship at Imperial College in London. Still, he yearned to establish physics in the Middle East and a few years later he helped set up the Institute of Theoretical Physics at Trieste. He is now head of the Institute.

In 1956 Salam attended a conference in Seattle at which Yang and Lee talked about the possibility that parity was not conserved in the weak interactions. The talk sparked Salam's interest. As he later said, "My trek to gauge theories began with that conference." In his Nobel address he told about the trip back to London from the conference in an American Air Force transport plane. There were so many crying children on the plane he couldn't sleep so he sat and thought about Yang and Lee's talk. Was it possible that parity was violated in the weak interactions? And if so why? A question his thesis advisor had asked him on his Ph.D. qualifying exam kept coming back to him: Why does the neutrino have no mass? Suddenly the answer came to him. "With a massless neutrino, parity had to be violated. It was the only choice nature had."

As soon as the plane landed in London the next morning, Salam rushed to his office and worked out the details of his idea. Who would he show them to? Who would know if the idea was right? His thesis advisor, Rudolf Peirls, the one who had asked him the question about the neutrino mass, would be the best bet. Excited and sure he had made a major breakthrough, he took his results to Peirls. But Yang and Lee had only conjectured that parity was not conserved by the weak interactions. It hadn't been proven yet—and Peirls was sure it was never going to be proven. He felt that parity was conserved and that Salam was wasting his time. Disgusted, Salam decided to try somebody else . . . Pauli. But Pauli wasn't at home when he visited him so he left his paper. Within a short time Pauli's reply came : "Salam . . . think of something better."

Salam was discouraged but he continued working on the theory. He soon teamed up with another theorist, John Ward, and the two of them published a series of papers. Like Glashow, Salam was also thinking in terms of a gauge theory. Basically, what he wanted to do was make Schwinger's theory into a gauge theory. The theory that they eventually developed was almost identical with Glashow's, yet interestingly they attacked the problem from an entirely different point of view. Like Glashow, they also found that their theory gave a neutral particle, and therefore neutral currents. Their major obstacle, however, was getting massive exchange particles. There seemed to be no way they could get them naturally, so like Glashow, they eventually gave up.

SYMMETRY BREAKING

Both Glashow and Salam finally decided that a major breakthrough was needed. Gauge theories were at an impasse. Some way had to be found to give the gauge particles mass without destroying the gauge invariance of the theory. Until now mass had been put in "by hand." Hints of the technique that would eventually be used came from Salam but he didn't develop them.

The breakthrough came primarily from Y. Nambu of the University of Chicago, and it was, at least in part, a result of luck. Nambu was an expert in both field theory and superconductivity, and during the mid 1960s was following developments in both fields. Spontaneous symmetry breaking had just caused a sensation in superconductivity, allowing the development of an extremely accurate theory of the phenomenon. Nambu soon realized that spontaneous symmetry breaking might be of value in field theory. And indeed it also eventually caused a sensation in field theory.

But what exactly is symmetry breaking? A simple example will, hopefully, give you a rough idea. Suppose a man is walking toward you; you know that the two sides of his body are symmetric. In particular, he has a right hand and a left hand that are nearly identical, but when he gets close to you he stretches out his right hand to greet you. What has happened? He has obviously broken the left–right symmetry.

Another example used by Salam is also illustrative. Suppose that a number of people are seated around a circular table. Between each of their plates is a napkin, so that each person has a napkin just to his left and just to his right. Everyone sits looking at the two napkins. Nobody is quite sure which one is his. There is an obvious symmetry. Suddenly someone reaches out and takes the napkin to his right. Everyone is then forced to take the napkin to their right, and the right–left symmetry is broken.

In the case of gauge theory it was believed that symmetry was also broken, or as some prefer to say: it was hidden. Hidden because it was not evident in the lowest, or ground state, but was evident at higher states (or, equivalently, higher temperatures). We can illustrate the situation by thinking of ourselves as living inside a large magnet. Let's not worry for now about the fact that magnets are solid. We know that a magnet has a magnetic field because millions of tiny magnets inside it are all lined up in the same direction. As you increase the temperature, though, these tiny magnets are agitated and it is difficult for them to remain aligned. And as a result we lose the magnetism. If the temperature inside the magnet were relatively high and you looked at your compass it would not point in any particular direction because there would be no overall magnetic field. This is a symmetric situation. As the temperature cooled, though, a magnetic field would form and your compass would point in a north–south direction. The symmetry would be broken.

Theorists were sure that broken symmetry was the key, but they had difficulty at first making it work. To make things worse a serious problem developed: Jeffrey Goldstone of Cambridge showed that when symmetry was broken in a field theory a particle always appeared. The problem wasn't the particle itself, but the fact that it was *massless*. We didn't need a massless particle, we needed a *massive* one. (The particle was eventually called the Goldstone boson.)

Was there any way this particle could become massive? Schwinger suggested that theorists should look closely at the difference between global symmetry breaking and local symmetry breaking. But for some reason he didn't bother to do the calculations himself. Fortunately, Phillip Anderson of Bell Labs picked up on his suggestion, arguing that there should be a way around the massless Goldstone boson because the corresponding particle in superconducting theory was not massless. But again few people paid attention to his paper. One who did was Peter Higgs of Kings College, England. Higgs had been following Nambu's and Goldstone's work and decided to see what would happen if the theory were made into a local one. To his surprise an exchange particle with mass appeared—a particle we now refer to as a Higgs boson. This was exactly what the doctor ordered. You would think that scientific journals would be ecstatic—practically lining up for the glory of publishing such a breakthrough. But they weren't—Higgs had trouble getting his first paper published. And his second one was rejected outright.

THE WEINBERG–SALAM THEORY

Steven Weinberg at MIT read Higgs's paper with passing interest. "At the time I regarded it as a technicality," he said. Weinberg was working on the strong interactions, and had developed a theory using the massless Goldstone boson. The pion was assumed to be the Goldstone boson, and the pion had a mass. But this didn't worry Weinberg. As he said in his Nobel address, "This may have been because of a new development in theoretical physics, which suddenly seemed to change the role of Goldstone bosons from that of unwanted intruders to that of

Steven Weinberg.

welcome friends." He therefore didn't take much interest in Higgs's discovery—at first.

Born in New York City in 1933, Weinberg received his undergraduate degree from Cornell in 1954 and his Ph.D. from Princeton in 1957. As a teenager he was a classmate of Glashow's at Bronx High School of Science. Both were members of the science fiction club. And it was during the meetings of this club that they learned most of the early physics. Long and sometimes heated discussions on the latest developments in physics took place here.

"I spent the years 1965 to 1967 happily developing the implications of spontaneous symmetry breaking for the strong interactions," said Weinberg in his Nobel address. Although Higgs's discovery, in itself, was of little initial interest to him, he did try to make the theory he was working on, a strong interaction theory, into a local one. It didn't work. Then, one day in the fall of 1967 it suddenly occurred to him that he "had been applying the right idea to the wrong problem." He should be applying spontaneous symmetry breaking and the Higgs mechanism to the weak interactions. He immediately went to work and soon had a theory that was so beautiful, logical, and elegant he could hardly believe it.

Putting the Higgs mechanism together with an $SU(2) \times U(1)$ theory of the weak and electromagnetic interactions he was able to produce the mass he wanted. There were a total of five particles in his theory, four of which had mass. He associated three of the massive particles with W^+, W^-, and the neutral particle Z^0. The massless particle was assumed to be the photon, and the last massive particle would be so massive it would not be visible. It was the Higgs boson.

But when Weinberg published his paper the typical reaction was "What else is new?" Nobody cared. The major reason for the lack of interest was that he had not shown that his theory was renormalizable. And so his paper laid unread for several years. At almost the same time Salam's interest in the problem was rekindled when Thomas Kibble, a theorist at Imperial College, told him about the Higgs mechanism. Salam saw immediately that the technique could be applied to the theory he and Ward had developed earlier. Like Weinberg he encountered a problem with renormalization and he also had neutral currents. Nevertheless, he was sure the theory was correct. He never published his theory but presented it briefly at a small conference in Sweden, and it was published in the proceedings. But to most of those present at the conference the significance was lost. Gell-Mann summarized the conference and never mentioned the contribution. So, just as Weinberg's paper lay buried for several years, so too did Salam's.

The next important step was taken by the Dutch theorist Martinus Veltman. Born in the Netherlands in 1931, Veltman studied physics at the University of Utrecht. Upon completion of his doctorate he became interested in renormalization of weak interaction theory, but soon found the problem was more difficult than he had anticipated. The higher-order Feynman diagrams all gave infinities, and if he was to solve the problem, he somehow had to show that they all canceled (in other words there were an equal number of plus infinities and minus infinities). He soon found that many of them did cancel—but not all of them. He tried several different methods, and eventually turned to the computer for help with the large number of calculations that were required. But nothing seemed to work.

Then he acquired a graduate student, Gerard 't Hooft. Upon completion of his bachelor's degree 't Hooft had decided he wanted to do his Ph.D. thesis in particle physics. Veltman was the only particle theorist at Utrecht so 't Hooft went to see him. Veltman suggested he take a class from him and write up the notes. During this time he could be thinking about what he wanted to work on. Veltman suggested several topics but none of them appealed to 't Hooft, who wanted a particularly hard problem, something that really challenged him. Veltman mentioned the gauge problem he was working on, but felt it was inappropriate. Almost no one was working on gauge theories, and to take it up would mean that 't Hooft would become a

Gerard 't Hooft.

specialist in an area nobody was interested in. Besides, the problem was so difficult that there was no assurance of success. After all, he (Veltman) hadn't been able to solve it. It seemed unlikely that a graduate student with little background in the area would succeed where he hadn't.

The problem sounded like the kind of challenge 't Hooft wanted and he decided to work on it. Veltman explained the details and off 't Hooft went. Some time later, after he had made some progress he returned to Veltman's office to show him what he had done. Veltman looked at it but was skeptical. 't Hooft seemed a little too cocky and Veltman was sure he didn't really understand the difficulty of the problem. After some effort, though, 't Hooft convinced him that what he had done was correct. He hadn't, however, completed the problem. Veltman asked about the last step—the renormalization. "I'll do it," he said confidently. And indeed he did.

When 't Hooft appeared with a method of demonstrating

renormalization Veltman was still skeptical but he now had a computer program into which 't Hooft's terms could be put. And when the output came he was shocked: all the infinities canceled. The theory was renormalized.

When Weinberg read 't Hooft's paper he wasn't convinced. The major problem was that it was written in a form that was unfamiliar to him. His friend B. T. Lee, however, was familiar with 't Hooft's techniques and soon translated it into something Weinberg could understand. This convinced him.

Suddenly there was a surge of interest in Weinberg's and Salam's papers. Everyone started digging them up. Then in 1973 neutral currents were discovered at CERN and the Weinberg–Salam theory was firmly established.

And since gauge invariance had been used in Weinberg–Salam theory, in other words in the unification of the electromagnetic and weak fields, it was obviously an important concept in physics. There was, however, still the strong field to deal with. Would gauge theory also be important here? In the next chapter we will see that it was.

CHAPTER 8

Adding Color

The quark model solved many of the problems of elementary particle physics. The hundreds of seemingly unrelated "elementary" particles and resonances were now understood. But searches were made for isolated quarks and none were found. Why? Perhaps they were forever confined to the interior of the hadrons as Gell-Mann had suggested. Many physicists were skeptical of the whole idea, yet the predictions of the theory were amazingly accurate. And they couldn't just be brushed aside. Still, there were problems—a particularly glaring one related to the spin of the particle. I talked earlier about spin, pointing out that the electron has spin $1/2$. Other particles such as the pion, on the other hand, have spin 0. We can, in fact, group particles into two classes according to their spin. The first class, those with half-integral spin (i.e., $1/2$, $3/2$, . . .), are called fermions; the second class, those with integral spin (0, 1, 2, . . .) are called bosons. Electrons, obviously, are fermions; pions are bosons.

Now, the problem. It was created because of a principle put forward in 1925 by Wolfgang Pauli. Pauli showed that when fermions are grouped together in a system, each of them has to be different in some way from all others in the system. Scientifically we say that they have to have different quantum numbers. This is now known as Pauli's Exclusion Principle, and as far as physicists know it applies to all fermions. And since we know that quarks are fermions it should apply to them.

But did it? It seemed not, for as we saw earlier the omega-minus particle had three strange quarks in it, each spinning in exactly the same way. In short, all three were identical, and this violated Pauli's Principle. Furthermore, it wasn't the only case. The delta particle (Δ^{++}) also consisted of three identical quarks. How could this be explained? There was, of course, the possibility that the quarks didn't obey Pauli's Principle; maybe they were, in reality, bosons. Bosons didn't obey the principle. But most theorists were reluctant to go along with this. After all, they had spin $1/2$ and the principle applied to all other fermions. Why wouldn't it also apply to quarks?

The solution to the problem came from Oscar Greenberg of the University of Maryland in 1964 in a very unexpected manner. Greenberg actually solved the problem before he knew there was a problem. At the time he had hardly heard of quarks, but he was interested in field theory; more particularly, he was interested in generalizing or extending field theories. And as one possible type of extension he invented a new type of statistics that he called "parastatistics." The problem with the invention was that it appeared at first that it wasn't needed. There were no field theories that needed extending. Then he became interested in quark theory and within a short time he saw that it didn't obey Pauli's Principle. He could, however, overcome this problem with his parastatistics. He was delighted. Working with A. Messiah he published a series of articles beginning in 1964. But there were difficulties and as a result his paper received little attention. The major difficulty was that there seemed to be no way to avoid what were referred to as parahadrons—hadrons with crazy statistics. Such particles couldn't possibly exist.

Greenberg's papers did, however, catch the attention of Yoichiro Nambu. About the same time Moo-Young Han, a graduate student at the University of Syracuse, wrote to Nambu showing him some work he had done on quarks. He suggested that SU(3) could be used as the basis of another charge (later called color). Nambu began an exchange with Han and between them they put together a paper. What they advocated, in modern terminology, was that each of the quarks came in three colors. They merely referred to it as a new type of charge, something like electrical charge, and this is, indeed, the best way to think of it. Although it is now called color, it has absolutely nothing to do with color as we know it. It is just a property of particles that is associated with the strong force.

This meant that we now had three flavors (types) of quarks (u, d, s) and each of these flavors came in three colors (say, red, blue, and green). The important thing, as far as Nambu and Han were concerned, was that physically observable particles had to be colorless. And since baryons were made up of three quarks, this meant that each of them has a different color, with the combination of the three colors adding up to white (being colorless). On the basis of this they were able to explain one of the outstanding problems of quark theory. Gell-Mann had advocated that baryons were made up of three quarks (qqq) and mesons of a quark and an antiquark (q$\bar{q}$). But why weren't other quark combinations such as qq$\bar{q}$, q$\bar{q}\bar{q}$, and qqqq important? Nambu showed that they weren't important (and didn't exist in nature) because there was no way such combinations could be made colorless. Only qqq and q$\bar{q}$ could be colorless.

Nambu worked on his theory for a few years after his paper was published but he eventually decided that the idea was too far-out. Certain that he wouldn't be able to get anything further on the subject published, he switched to other problems. The colored quark theory languished for a while but it didn't die. It was, in fact, suddenly and surely brought to life by the discovery in the early 1970s that it could explain two important phenomena. The first was the decay of the neutral pion to two photons. When a calculation of this process using ordinary quark theory was made it was discovered that the prediction was off by a factor of nine. But then it was shown that the prediction actually depended on the square of the number of quark colors. And if there were three colors there was almost exact agreement with experiment.

The second process that was explained by color theory was one in which an electron interacts with a positron. If the energy is sufficiently high mostly hadrons are produced; at lower energies, on the other hand, most of the particles that are produced are muon–antimuon pairs. When theorists calculated the ratio of the frequency of these two reactions, in other words, the number of hadrons divided by the number of muon pairs, they found they were off. But when colored quarks were included theory came in line with experiment.

FREEDOM AND SLAVERY

There are few aspects of quark theory that Gell-Mann did not have a hand in, and colored quark theory was no exception. Although he did not invent the color concept he made numerous contributions to the theory and was a strong and ardent supporter of it when few others were interested. In 1972 he teamed up with the German theorist Harald Fritzsch, and the two of them began working on an extension of Nambu's ideas. The theory that they developed was different from Nambu and Han's in two important respects. First, Nambu had decided that fractionally charged quarks were not needed; he replaced them with integral charged quarks. Furthermore, his theory was not locally gauge invariant; in other words it was not a Yang–Mills theory. Gell-Mann preferred his fractional charges; furthermore, he was convinced that gauge invariance was an important aspect of any theory of the strong interactions and he incorporated it into his theory.

Most of the terminology we use today came from Gell-Mann. Nambu had merely referred to a new type of charge; Gell-Mann called this new charge, color. And since he was modeling his theory on quantum electrodynamics it seemed natural to call it quantum chromodynamics (QCD), the "chromo" coming from the word "color." The new theory was an important step, but like his old theory, it too had problems. The major one was that it could not explain scaling. Scaling, you may remember, was explained by Feynman through his introduction of partons. But quarks—even colored quarks—behaved quite differently from partons. The partons in the proton acted like free particles. Colored quarks, on the other hand, acted no different from ordinary uncolored ones. There was nothing in the theory that said that they were free at short distances. Another problem with the new theory was one that had plagued the old theory—confinement. Why and how were the quarks confined inside the proton?

The confinement problem is still with us, but we have resolved the conflict between the parton and quark models. The first step came from Ken Wilson. Wilson majored in mathematics at Harvard but upon graduation in 1956 he decided he wanted to switch into physics, so he went to Caltech to work under Gell-Mann. Upon finishing his course work he went to Gell-Mann, asking him about a thesis project. Gell-Mann suggested several problems, but none of them appealed to him. He wanted something "long term," something that would take him a while to get results, and would be a substantial contribution to physics. Gell-Mann suggested that he look at a field theory interpretation of the strong interactions. But he warned him that it was not an "in" subject. Most people had given up on such an attempt. Wilson was interested, the subject appealed to him, and he began working on it, eventually completing his Ph.D. in the area. Upon graduation he returned for a while to Harvard, then in 1963 he went to Cornell. All the while he continued working on the strong interactions, trying to apply various field theory techniques. The work progressed slowly; he barely managed to publish a paper every two years for the first ten years after he graduated. But he did make progress.

He became interested in a technique called the renormalization group that had been invented by Gell-Mann and Francis Low in 1954. Some of the ideas of the method had been developed several years earlier by Ernst Stueckelberg and his student André Petermann, but it is the Gell-Mann–Low paper that has

David Politzer.

become the standard reference. It was an interesting but little-used technique that allowed you to make predictions at one energy level if you knew what happened at a different energy level. Wilson worked with the renormalization group technique and showed that it could be used fruitfully in the strong interactions, but he did not make the major breakthrough that the technique eventually brought.

One of those who did was David Politzer of Harvard University. In early 1973 Politzer was a graduate student; he had finished his preliminary exams and was working on his thesis but things weren't going well. He began to worry that he would not be able to get enough data to complete his thesis. Then he learned about the renormalization group technique and became interested in it. It was a much more fruitful area for a thesis than the one he was working in. His first idea was to apply it to electroweak theory, but when this didn't work he turned to the strong interactions. Maybe he could use it to explain spontaneous symmetry breaking. It seemed like a good problem so he went to his thesis advisor and asked him if anyone had done it.

His thesis advisor, Sidney Coleman, who was on sabbatical at Princeton University at the time, didn't know if the problem had been solved but was sure one of the Princeton faculty, David Gross, would know. So the two of them went to Gross's office and asked him. And indeed Politzer got the answer he hoped for: no one had solved the problem. He therefore immediately began working on it.

Oddly enough, Gross was working on a similar problem. He too was interested in the renormalization group and wanted to use it to explain scaling. To do this he had to find a field theory that would predict scaling. He had tried all the easy theories, and found that none of them worked. At this point he had became frustrated—it almost seemed as if nothing was going to work. The only theories left to try were Yang–Mills type theories, and they were difficult to work with. About this time a graduate student, Frank Wilczek, came to work under him. Gross decided that an investigation of these theories would be a good thesis project for him, so he assigned it to him. Wilczek had barely got started, though, when he got a scare. Upon reading a paper by the European theorist Kurt Symanzik he discovered that Symanzik was also interested in the problem. Afraid he might get scooped on his thesis he shifted into high gear. But problems plagued him. He began by showing that Yang–Mills theories could not explain scaling, as Gross had hoped, but then discovered he had made a mistake. He corrected it, only to discover another mistake, which he also corrected.

Meanwhile, Politzer, at Harvard, was working on basically the same problem, but from a different point of view. To his delight he discovered that Yang–Mills theories were "asymptotically free." This meant that the quarks inside hadrons would act like free particles over short distances. He was sure it was an important discovery and took it to Coleman in anticipation. "Sorry," said Coleman, "you can't be right. I just talked to Gross, and his student, Frank Wilczek, has proven that Yang–Mills theories are not asymptotically free. You must have made a sign error." Politzer was stunned. But it was possible: the calculations were long and he could have made an error. He carefully checked all his calculations; in fact, he checked them several times and each time they came out the same. Now there was no doubt in his mind: he hadn't made a mistake. There was no sign error. So back to Coleman he went to tell him. By this time Wilczek had found his mistake and word of it had reached Coleman. Politzer was, to say the least, relieved. Not only did he now have enough material for a thesis but he had made an important discovery: Yang–Mills theories were, indeed, asymptotically free. Papers on the phenomenon were published by both the Harvard and Princeton groups in the same issue of *Physical Review* in 1973.

Interestingly, the discovery had already been made across the Atlantic in Holland. Gerard 't Hooft had looked into the problem and also found that Yang–Mills theories were asymptotically free. But when he told Symanzik about it, Symanzik shook his head and told him he had just made a sign error. Undeterred, 't Hooft announced the discovery at a meeting but he did not publish it and failed to follow up on it. Apparently he did not realize its importance.

Gell-Mann and Fritzsch were delighted when they heard about asymptotic freedom. With asymptotic freedom added to their model everything fit. One of the important properties of their theory was that it was locally gauge invariant. And as we saw earlier whenever a global theory is made into a local gauge theory a new field is created. In this case the new field would be the interaction field between the quarks. Gell-Mann called the field particles, gluons. They were, in essence, the "glue" that held the quarks together. As the quarks move around inside the proton they exchange gluons. Because of asymptotic freedom, when they are close together hardly any gluons are exchanged, but as they separate the number increases.

These gluons are like the photons of quantum electrodynamics, but they are much more complex. To begin with, there are eight types of gluons versus only one type of photon. But there is an even more important difference: photons are neutral and therefore don't react with one another; this isn't the case for gluons. Gluons are colored and therefore react among themselves. In short, they "stick" to one another. This is an important difference and it has important implications.

Now let's go back to asymptotic freedom and take a closer look at it. It occurs because the quarks, when close together inside, say, the proton, are able to move around freely. They are not bound together, and feel essentially no force. This is, of course, what Feynman found for his partons. This meant that the quark model was no longer in conflict with the parton model. In short, partons were quarks.

When quarks are close together, then, they act like free particles, but as the distance between them increases the force between them increases. And eventually they feel an exceedingly strong force. The force I'm referring to here is, of course, the color force. Compared to the electrostatic force, then, it is quite different. Where the electrostatic force decreases with distance (between charges) the color force increases.

How can we explain this seemingly strange phenomenon? The best way, perhaps, is to take a closer look at what is going on in the neighborhood of the quark. We'll begin by briefly reviewing the nature of the charge cloud around the electron. We saw earlier that virtual pairs can be created in the neighborhood of an electron. The electrons created within these pairs will be repelled by the central electron, but the positrons will be attracted. Thus, the cloud will break into two distinct regions: an outer generally negative region, and an inner positive region. The overall result is a shielding of the true charge of the electron. We refer to this true charge as the "naked charge." Since we are at a considerable distance from the electron we see only what is called the physical charge, in other words the naked charge minus the shielding. Because of this we find that the force does not fall off (decrease) the way it should close to the electron.

A quark will also have a cloud around it. It is composed of quarks and antiquarks, but there are also gluons present since they are passed back and forth continuously between quarks. Furthermore, there is a polarization of the gluon cloud. But in this case there is not a shielding of the color charge as there was in the case of electron–positron pairs, but an enhancement of it. As you get farther away from the charge the attraction becomes greater instead of falling off. This means that when a quark ventures too far away from the other quarks in, say, the proton, it is pulled back. We sometimes refer to this as "slavery." The quarks are generally free as long as they don't stray too far, but when they do, they aren't able to go far.

QCD

Quantum chromodynamics (QCD) is now considered to be a highly accurate theory. And just as quantum electrodynamics is the theory of electromagnetic interactions, quantum chromodynamics is the theory of the strong interactions. Like the older quark theory it is based on the group SU(3). The "triplet" of quark theory is, in fact, now considered to be a color triplet (red, green, and blue) rather than a flavor triplet (up, down, and strange). Another difference is that the new theory is exact. The old theory was based on the up, down, and strange quarks, which we know have different masses, so it was not an exact symmetry. The new theory is based on the three colors—red, green, and blue. And quarks of the same type, but of different color, have the same mass.

Another important property of quantum chromodynamics is that it explains why only quark–antiquark pairs and triplets of quarks exist. They are the only colorless combinations. Furthermore, as we just saw, the color force is assumed to be the result of exchange particles called gluons. And just as quarks are colored, so too are gluons. In fact, gluons have two colors associated with them. Since there are three basic colors and each gluon has two color labels it would appear as if there should be nine kinds of gluons (i.e., $3 \times 3 = 9$). One of them, however, turns out to be colorless and as a result there are only eight different types.

It is also worth noting that the uncertainty principle is in effect here as it is in the case of other atomic phenomena. We cannot say for certain that a particular quark in, say, a proton is red. It only has a certain probability of being red, or perhaps a better way of saying this is that it is red "on the average." The important thing is that "on the average" there is one of each color so that in total they are colorless.

I use the words "on the average" because quarks are continually changing color. When a quark absorbs or emits a gluon it changes color. If a red quark, for example, emits a red-yellow gluon it changes to blue. On the other hand, if a blue quark absorbs a red-yellow gluon it changes to red. In short, then, gluons, or perhaps I should say, the strong interactions, can change the color of a quark. They cannot, however, change a quark's flavor. If the quark was originally an "up" quark there is no way the strong interactions can change it to a "down" quark. A change in flavor can, however, be accomplished by the weak interactions. When a quark emits or absorbs a W particle its flavor changes, but its color remains the same.

Because gluons interact with one another there is also the possibility of an entirely new type of bound state—two gluons bound together as a "glueball." This glueball would have to be colorless so the two colors of one of the gluons would have to cancel the two colors of the other (e.g., red–antiblue + blue–antired).

How would we generate glueballs? One way would be in an electron–positron collision. Its lifetime would be extremely

A *glueball*.

short, on the order of 10^{-25} second so it would not be possible to observe it directly. But it could be detected indirectly by observing the particles that are produced in its decay. Experimenters at Brookhaven believe they have detected one, but confirmation is still needed.

If you think about gluons and the strong nuclear force for a moment it may seem like we have a problem. The strong nuclear force we observe is not between quarks, it is between nucleons (protons and neutrons). How does the color force relate to the nuclear force? To answer this it is best to begin with atoms. We know the electrons are held in place by the electromagnetic force. But atoms are, of course, also held together as molecules, which results from an overlapping of the electron wave clouds (the so-called van der Waals forces). In the same way the strong nuclear force is the van der Waals force associated with the color force.

IMPRISONMENT

As I have said many times in the past, one of the major problems of modern particle theory is the problem of confinement. Are the quarks necessarily confined to the interiors of hadrons? Or is it possible that some day, with enough energy, we will be able to pull one out? Most physicists believe they are permanently confined, but so far no one has been able to prove

Electric field lines around two charges that are relatively close to one another.

it. We know, of course, because of the slavery phenomenon discussed above, that it is difficult to pull one very far. The harder you pull the greater the force with which the quark is pulled back. We might ask: would it be possible to pull a quark say, one meter out of the nucleus? To answer this let's begin with the analogous case of two electrical charges. We can represent the force between them as lines of electrical force. The more lines in a given volume, the greater is the force. If we increase the distance between the charges the lines of force curve up and become less dense as shown in the figure. This, of course, corresponds to Coulomb's law, which says that the force between the two charges decreases as the distance between them increases.

Now let's consider two quarks. We can also represent the color force between them by lines of force. We refer to them as chromoelectric field lines. But it turns out that there is a major difference in this case. Associated with the photon is an electric and magnetic field, so it is reasonable that we would have a similar situation in the case of the quark. And we do. The two types of lines are known as chromoelectric and chromomagnetic lines. In the case of the photon there is no self-interaction so we

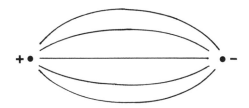

Electric field lines around two charges that are farther apart.

don't have to worry about the two fields interacting. But gluons do interact with one another so we get an interaction. A simple representation of it is shown below.

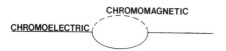

A chromoelectric line splitting into a chromoelectric and a chromomagnetic line.

What this means is that there will be an interaction between the chromoelectric lines that stretch between the two quarks. Initially, when the two quarks are relatively close we will have the lines of force looking as in the figure below.

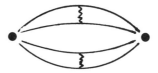

Lines of force between two quarks.

There will, in effect, be almost no interaction between the chromoelectric lines. But as we pull the quarks apart the interaction (chromomagnetic) lines will increase and draw the chromoelectric lines together. And since the number of lines per unit volume is a measure of the force, the force between the quarks will increase. Eventually the lines will become parallel and they will look almost like a "string" between the quarks.

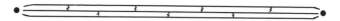

Lines of force between two quarks when quarks are a long distance apart.

When does this happen? We don't know for certain but many believe it has occurred by the time the quarks are 10^{-13} centimeter apart (this corresponds to the size of the proton). From there on the force would presumably stay constant, but it is already so great at this point that it would take more energy than any accelerator on Earth could supply, to separate them any farther.

Long before this happens, though, the string breaks and we get two quarks, or I should say, a quark and an antiquark, which we know is a meson. Thus, if we bombard a proton, hoping to knock one of the quarks out, we end up getting a meson.

Nambu and others picked up on the string idea and developed it. According to Nambu the quarks are attached to one

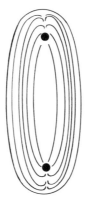

Bag model of confinement. Lines shown are field lines between quarks.

another by strings. For simplicity we can think of them as elastic strands. Nambu thought of them as made up of gluons. As long as the quarks are relatively close to one another the elastic strands are loose and there is little force between them, but when the distance between them increases the elastic tightens and begins to pull back. The farther they move away, the greater is the force. Nambu eventually had problems with his theory but, interestingly, it is now the basis of a modern string theory we will talk about later.

Another model of confinement is due to Ken Wilson. It is referred to as the bag model. In this model the quarks of an individual hadron are confined in a bag which is kept inflated by the pressure of the quarks inside it. They are, in a sense, like the gas molecules in a rubber balloon. The quarks cannot penetrate the bag and neither can their color lines of force. Again, if we try to pull the quarks apart we elongate the bag and squeeze the color lines together so that they become stronger. Eventually, with enough of a pull, they become parallel and our elongated bag resembles the string we talked about earlier.

Each of the above models explains certain aspects of confinement, but none of them explains everything. Theorists strongly believe that quantum chromodynamics will eventually explain confinement but a complete and satisfactory explanation of it is still not at hand. But as we have seen, scientists were finally beginning to understand the strong interactions. Then came charm.

CHAPTER 9
Adding Charm

The story of charm—a quality embodied in a fourth quark—begins not with charm itself, but with neutral currents. We talked about such currents earlier. They occur in neutrino reactions. Most neutrino reactions involve "charged currents"; in these reactions the neutrino changes into an electron and in the process there is a transfer of charge. But according to the Weinberg–Salam theory there should also be cases where there is no transfer of charge. This is the neutral current case. When a neutrino goes into a reaction in this case a neutrino comes out. The neutral current is associated with the neutral boson Z^0 that passes between the two particles (see Chapter 7, page 137).

A number of theorists were certain that neutral currents existed but as late as 1970 they had still failed to pass on their enthusiasm to experimentalists. Steven Weinberg was one of those who was particularly interested, primarily because the theory he had formulated was at stake. If neutral currents could not be shown to exist, it would go the way of all interesting—but incorrect—theories. Weinberg's cause got a shot in the arm in 1971 when 't Hooft showed that the Weinberg–Salam theory was renormalizable. This brought it to the attention of experimentalists and a few of them began to wonder if it was perhaps worthwhile to begin a search. Most, however, were still skeptical.

In 1971 Weinberg went to Fermilab to encourage experimentalists to initiate a search. At about the same time theorists in

Europe began pressing experimentalists at CERN. To many, CERN was the best bet. A giant bubble chamber called Gargemelle, designed specifically to detect neutrinos, was being built. Until now it had been extremely difficult to detect neutrinos; most neutrinos striking the Earth would go right through it without interacting. It was therefore obviously difficult to get them to interact with the particles of a tank that was infinitesimally small compared to the Earth.

But Gargemelle was up to the task. It weighed a thousand tons and contained ten tons of Freon. (The strange name came from a 16th century tale by Francois Rabelais. Gargemelle was a gluttonous giant who gave birth to another glutton, Gargantua, after stuffing herself with hogheads.) Of the billions of neutrinos that would enter Gargemelle, most would pass completely through, but occasionally—very occasionally—one would react with a Freon molecule and betray its presence. But there was a further complication: the track of the neutrino would be invisible. They are neutral and therefore do not leave a track as charged particles do. Their positions would therefore have to be determined from the other particles involved in the reaction.

By the time Gargemelle was ready for its first run there was an air of anticipation. Neutral currents might, for the first time, be "seen." But the experiment would have to wait its turn. There were teams from almost every European nation, and many other countries, including the United States, also wanting to use the giant. Weak neutral currents were not a priority.

The director of the Gargemelle project, Paul Musset, heard of neutral currents while he was directing the construction of the giant bubble chamber. But he was far too busy to take an interest. Then his job was over and he finally had time to think about physics. The mounting interest in neutral currents caught his fancy and he decided to see if he could find any evidence of them. After persuading several collegues to assist him he began to consider what he would look for. He decided that the presence of an electron coupled with the absence of a muon would be the most interesting event. Because of the strong magnetic field in the chamber the electron would be easy to spot: its track would be a tiny spiral. But the muon would be more trouble because there was the danger of mistaking its track for that of a pion. Nevertheless they had a goal. And after most others had left the laboratory, Musset and his colleagues went through photograph after photograph. After checking literally hundreds of thousands of events they finally came upon an interesting one. Musset was ecstatic. But a single photograph was not enough. Musset decided the best thing to do was to use it to interest others—which he did. His enthusiasm was now brimming over and he became more determined than ever: he would be the first to discover neutral currents.

But unknown to him an American at the same laboratory was searching for the same prize. Robert Palmer had gone to CERN in 1971 to work on Gargemelle. But when he arrived he found the apparatus he was to work with was not ready so he decided to look around for another project while he waited. Neutral currents filled the bill. It would not involve experimentation; he would be able to use data that had already been accumulated. So, like Musset, he persuaded a couple of colleagues to help him, and he began the search. And like Musset he also soon found evidence that convinced him that neutral currents existed. He showed it to several experimenters in hopes of interesting them, but to no avail. No one took him seriously. The method he had used was crude, and it was quite possible that the events could be due to something else.

Palmer was put off by his inability to spark anyone else's interest, but he was convinced that it was an important discovery and he wasn't going to just shrug it off. He wanted to publish, but there was a problem: he had used data that had been taken by other groups.

Meanwhile, where Palmer had failed, Musset had succeeded. He had gotten others interested in his evidence. In fact, he managed, indirectly, to get the one person who counted the most interested: the director of CERN. The search was soon given high priority. Many others joined in, but success did not come immediately. Finally, though, in early 1973 another interesting event was found, and Musset began talking about the discovery at various international meetings.

Palmer, who was now back in the United States, attended one of these meetings and was shocked when he heard of Musset's discovery. But a public announcement had not yet been made. There was still time—if he got into print fast enough. He obviously couldn't publish an experimental paper—that would be unethical. But there was an alternative: he could publish a theoretical paper that incorporated the results. He quickly wrote up the paper and submitted it to *Physics Letters*. But *Physics Letters* was published by CERN, and by now the editors and referees had heard of the interest throughout CERN. They obviously weren't going to let an American scoop them. So the paper sat.

Finally CERN officials, including Musset, were satisfied that there was enough evidence for an announcement, and they went public in July 1973. Shortly thereafter a paper presenting their result appeared in *Physics Letters*. Palmer's paper was published a few weeks later in the same journal.

Musset knew nothing of the discovery of Palmer and his group—at least not until Palmer attempted to publish. His worries were elsewhere: he knew the Americans were not sitting idly by. At Fermilab a similar search was under way. A group led by Carlo Rubbia were also hot on the trail of the elusive current. Although they were plagued by troubles for months, Rubbia and his group finally obtained some interesting results. They wanted to publish, but had to admit to themselves that they were uncertain of their interpretation. One of the major difficulties was that several groups had shown that neutral currents never occurred in the decay of strange particles. Yet there were indications that they should. Maybe what they were seeing were not neutral currents after all. They were reluctant to take a chance and as a result lost the honor of being first.

Rubbia's group was discouraged when the announcement

came from CERN, but they continued to wait. Finally, though, it became evident that what they had seen were neutral currents and they published. With the discovery at CERN and the verification at Fermilab there was now no doubt: neutral currents existed. And this was exactly what was needed to verify the Weinberg–Salam theory.

But where does charm come into this? So far we haven't mentioned it. To answer that we have to go back to why neutral currents were not found in strange particle decays. Several years earlier Glashow, John Iliopoulos of Greece, and Luciano Maiani of Rome had published a paper showing that neutral currents would not be produced in strange particle reactions. Their reasoning was based on the assumption that there was an as yet undiscovered quark—one called charm. Since their prediction had been borne out, charm must exist.

CHARM

Oddly enough, neither of the two teams that eventually found charm were actually looking for it. Verification of Glashow–Iliopoulos–Maiani theory was not their motive.

One of the two teams was led by Sam Ting. Ting was born at Ann Arbor, Michigan in 1936 while his parents were visiting the University of Michigan. Soon after his parents returned to mainland China they had to flee to Taiwan because of the war with the communists. Ting received most of his early schooling in Taiwan but had a desire to come to the United States, and upon graduation from high school in 1956 he decided to return. Following in his father's footsteps he entered the University of Michigan and began working on a bachelor's degree. By 1962 he had his Ph.D. After teaching for a while at Columbia University he went to MIT.

Ting is a quiet, soft-spoken man, but beneath the apparent calm exterior is an incredible intensity. He is fiercely ambitious and at times works both himself and those under him almost to exhaustion. He has a strong aversion to error and is meticulous in all aspects of his research. Where others check their results once or twice, he checks them numerous times. And even then he is frequently not satisfied. He always has at least two teams processing data independently. To many, such excesses are a waste of time—but not to Ting.

Ting gained considerable expertise in dealing with electron–positron pairs while working at DESY in Hamburg, Germany. When he returned to the United States in 1972 he wanted to put his newfound knowledge to use. He was particularly interested in the possible existence of what he called "heavy photons." Three such particles were known to exist, called rho, phi, and omega, and Ting was sure that there were similar but heavier particles. Along with Min Chen of MIT he wrote a proposal to Fermilab. With its huge new accelerator Fermilab was a natural choice, and Ting was certain he would make good use of the new instrument. But Fermilab officials didn't share his confidence; they were sure heavy photons didn't exist. Ting would be wasting his time. And so they made excuse after excuse as they delayed his proposal. Ting dealt with them calmly at first, but then he began to get frustrated.

Disgusted with Fermilab, Ting turned to CERN, but he also got little encouragement there. He continued traveling back and forth between Europe and America, talking with officials, pleading with them. But to no avail. Finally he decided he had had enough. He withdrew his proposal from Fermilab and early in 1972 submitted one to Brookhaven National Laboratory. A few months later it was approved.

Ting was overjoyed and was soon hard at work designing and building equipment. One of the major pieces of equipment that had to be built was a detector. In the experiment, high-energy protons would hit a target, creating many types of particles. Among the debris would be electron–positron pairs, and it was these pairs that Ting was interested in. The debris itself was unimportant, but it was a nuisance because the pairs would have to be distinguished from it. Since the detector would do the distinguishing, it had to be exceedingly sensitive.

Ting was not satisfied with it just doing the job. To many he seemed to go overboard in his zeal for sensitivity and accuracy. Finally, though, in the spring of 1974 everything was ready for the first runs. After checking out the equipment the crew began scanning the energy spectrum. If there was, indeed, a new heavy photon at any of these energies, its location would be signaled by a sudden upsurge of electron–positron pairs, which, in turn, would give a peak in the energy spectrum. So as they eased their way through the spectrum they carefully counted the pairs. On the first run they found nothing. Then they had to turn the beam over to another group and wait.

The waiting was a nuisance to Ting but not as frustrating as the delays and breakdowns that occurred when he finally got back on line. He moved now to lower energies. Always cautious, Ting had two teams independently analyzing the results. One team was headed by Ulrich Becker, the other by Chen. Each had to process the data on the computer; what came out was a listing of the number of electron pairs at various energies. In early September 1974 the first data from the second run were ready, and the two teams began processing it.

Shock came over Becker as he saw the first plots. The points were all piled up in the same region around 3.1 billion electron volts (3.1 GeV). He was sure it had to be a mistake. The peak was like a needle—sharper than anything he had ever seen. Was the equipment faulty? Or perhaps the computer program? He went through and checked the program and data carefully. The peak remained.

Chen, the leader of the second data analysis team, expressed similar concern when he encountered the peak. He couldn't believe it. He wondered if it would show up on Becker's results. If it didn't he would be embarrassed to report it to Ting. Still unsure of himself he phoned Ting and told him the news. Ting immediately called a meeting of the group.

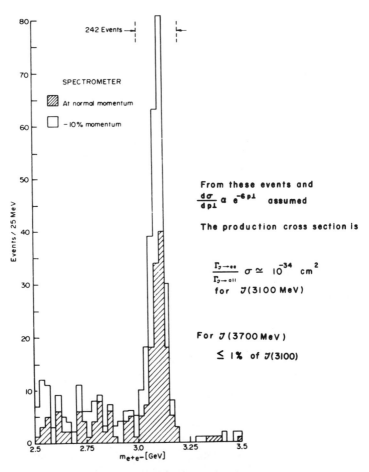

A graphical representation of the J/ψ particle showing peak at 3.1 GeV. (Courtesy Brookhaven.)

Ting and his group at Brookhaven. (Courtesy Brookhaven.)

The spectrometer used in the discovery of the J/ψ particle. (Courtesy Brookhaven.)

Chen and Becker were relieved to find that they had both found the peak. But this still didn't rule out equipment malfunction. A check on the peak would have to be made. They would definitely have to verify it before they could publish. But there was a problem: their time on the accelerator was up and it might be two months before they could get back on it.

Ting was uncertain what to do. But he was certain about one thing: secrecy was of the utmost importance. He gave strict orders that no one was to say a word about the discovery. And the reason was obvious—particularly to Ting. For he knew that across the continent at SLAC a similar experiment was going on under Burton Richter. The experiment was different from Ting's in some respects but if Richter and his group heard that there was a peak at 3.1 GeV they could easily find it—probably within a day. Ting wondered if they were working in the region near 3.1 GeV. He was relieved when he heard indirectly that they had passed through the region and had found nothing. This gave him breathing time; he would be able to check, hopefully, before Richter found it.

THE SPEAR EXPERIMENT

Richter's group at SPEAR had indeed passed 3.1 GeV without finding the peak. It was exceedingly narrow and their detection apparatus was not as sensitive as Ting's.

Richter is the antithesis of Ting. Easy-going and friendly, he frequently jokes and laughs with the men who work under him. Born in Brooklyn in 1931, he became interested in science as a youth after receiving a microscope as a present. When he was a teenager he built a chemistry laboratory in his basement and considered going into chemistry. But at college, after taking classes in both chemistry and physics, he decided physics was more to his liking. In 1955 he obtained his Ph.D. from MIT, then went on to Stanford. And he has been at Stanford ever since.

His early career at Stanford was, in many ways, frustrating. Soon after coming to Stanford he became interested in a technique invented by Gerard O'Neill of Princeton in which electrons and positrons were "stored" by circulating them at a speed near that of light in large rings. His dream was to build such a facility at SLAC, but it was over 20 years before the dream was finally realized.

The new facility was called SPEAR. Electrons and positrons from the two-mile-long linear accelerator were introduced and stored in the SPEAR rings. They entered as needle-shaped bunches an inch or so long and circulated in opposite directions.

The bunches passed one another millions of times each minute without interacting. But occasionally an electron would hit a positron and annihilation would occur. The energy that resulted would produce new heavy particles. Richter's experiment was, basically, the opposite of Ting's. He started with electron pairs and caused them to collide, then analyzed the results. Ting started with protons colliding with other protons and created electron pairs.

Richter's technique had several advantages. First, everything that was created came out of the electron–positron annihilation. In Ting's case the proton–proton collision created a substantial amount of debris besides the wanted pairs, and the

The storage ring at SLAC. (Courtesy SLAC.)

The Mark III detector at SPEAR.

pairs had to be distinguished from this debris. Furthermore, in Richter's experiment all of the energy of the particles went into the making of new particles (the particles are moving in opposite directions and they have the same mass). In Ting's experiment, protons were striking other protons in a stationary target, and most of the energy, therefore, did not go into producing new particles.

Unlike Ting, Richter was not looking for new particles—at least not directly. He was interested in a ratio referred to as R. We talked about it earlier; it is the ratio of the number of electron–positron collisions that produce hadrons to the number that produce muons. This number was, as we saw earlier, one of the first important tests of quantum chromodynamics. Experi-

CHAPTER 9

Burton Richter.

mentally it had been shown to have a value of 2, and this was in agreement with a quark model in which there were three quark colors. Theorists were pleased with the result, but when data were taken in 1973 at near 5 GeV R appeared to have risen substantially. There were indications it was near 6—three times the expected value. Theorists began to worry, but the result would have to be verified before it would be accepted.

Richter was interested in verifying this preliminary result. And, indeed, he soon did. R appeared to rise linearly with energy. The question that now confronted him was: would it level out? Or would it continue to rise indefinitely?

What, you might ask, would cause such a rise? It was well-known that the generation of new particles at a particular energy would cause a bump in the data. And in the preliminary runs Richter's group noticed slight bumps at approximately 3 GeV and 4 GeV. Both were worth checking so they decided to check the 4 GeV one first. They were working in the region when one of the group, Roy Schwitters, decided that a preliminary report of the results was in order. He started to write a paper, but then decided it was best to wait until the 3 GeV bump was checked. The check was made. But again the sharp spike at 3.1 GeV was missed.

Meanwhile Ting finally got back on line at Brookhaven. The check was positive: the peak was still there. But further tests were required. Changes were made in the target, in the beam and various other parts of the apparatus. Still the peak remained. There was no doubt now: it was a real peak. They could publish. But Ting was reluctant. If there was one peak in the region there must be more; if he published now everyone would immediately jump on the bandwagon and begin searching for them. And they might find them before he did. He therefore decided to delay while he checked the surrounding spectrum for further peaks.

Chen panicked when he heard Ting's decision. He was sure they were going to be scooped. By now a few others knew of the discovery and everyone encouraged Ting to publish. But he stubbornly refused.

While Ting was hedging, Richter was getting ever closer to the discovery. There was a sudden interest within his group in the 3 GeV bump. One of the members of the group had begun talking about charm, speculating that the bump might be associated with it. Several members pressed for a thorough recheck. But they were now at higher energies and Richter was reluctant to make the required changes. But the pressure on him became so great that in early November he finally gave them the go-ahead.

Carefully they squeezed in on the small bump near 3 GeV. At first few pairs were generated but then the number began to increase. Soon it was 30 times background and the group knew they had something. Everyone huddled over the instruments as they squeezed in closer. Finally the peak was 70 times background. Champagne bottles were broken open. Everyone was in a state of ecstasy; some of them actually began writing a paper announcing the event.

While this was going on Ting was on his way to SLAC for a meeting. Chen had pleaded with him at the last minute to publish but he stuck to his guns. Little did he know how close he was to being scooped (and if it were not for the meeting he might have been). Arriving in San Fransisco on Sunday, November 10, he went directly to his hotel room where he received a telephone call. He was stunned as he heard the news: Richter had discovered a new particle at 3.1 GeV. But he was determined not to be scooped. He quickly phoned Europe and announced his discovery.

The next morning he met with Richter. "I have some great news to tell you Sam," Richter said as he entered the room. Before he could continue Ting countered with, "And I have some great news to tell you." The two men told each other of their discoveries. Ting had called the particle J; Richter had called it psi (ψ). As a compomise we now call it J/ψ.

News of the discovery was phoned to laboratories around the world. Early in December three papers appeared in *Physical Review Letters*: one by Ting and his group, one by Richter and his group, and a third one from an Italian laboratory that had been doing research in the same area. It was a monumental discovery, and within a year Ting and Richter were awarded the Nobel prize.

Since Ting made his discovery several months before Richter it is natural to ask if perhaps word somehow leaked out. If anyone in Richter's group had any indication that someone had found something at 3.1 GeV they could have quickly checked on it. Ting is of the belief that this is possible, but Richter vehemently denies it. We will, perhaps, never know. But from all indications the discoveries were completely independent. Richter expressed genuine surprise when he heard of Ting's discovery. And of course his reasons for looking at the slight bump around 3 GeV are valid. Anyone would have done the same thing under the same circumstances. Besides, Ting had no proof that anyone in his group had leaked the news.

Another controversy regarding the discovery appears in relation to the awarding of the Nobel prizes. There were many people involved in the research, each contributing a part. In fact, some of the senior members of each group were quite well known in their own right. Yet only the two leaders got the prize. Some have argued that this is not right. It will no doubt have some effect in the future when it comes to selecting group leaders.

PROPERTIES OF THE J/ψ

The new particle (J/ψ), which was also soon referred to as charmonium, was discovered at an energy of 3.1 GeV, and it therefore has a mass of 3.1 GeV. This is three times the mass of the proton. It was, at the time, the heaviest particle known. It had no charge and was soon shown to be a meson. This meant that it was made up of a quark and an antiquark. The excitement, of course, centered around the fact that all quark–antiquark combinations had been accounted for. It could not therefore be made up of u, d, or s quarks. It had to be made up of a new type of quark.

Its major distinctive property was its lifetime. Lifetimes of similar particles, assuming they decayed via the strong interactions, were of the order of 10^{-23} second. This particle lived a thousand times longer than that. How do we know this? The lifetime of a particle, it turns out, can be determined from the width of the peak. If the peak is relatively broad the particle is short-lived; if it is narrow, as the J/ψ peak was, the particle is long-lived.

But what exactly was the J/ψ? Within days after the discovery most of the speculation centered around charm. Glashow and others had been talking about it for years. But the properties of the J/ψ indicated that it couldn't be a single charmed quark; it had to be a combination of a charmed quark and an anti charmed quark (c$\bar{\text{c}}$). It was, therefore, a particle with "hidden charm." Such particles had been predicted by Mary Gaillard, Benjamin Lee, and Jonathan Rosner.

THE CHARMONIUM SPECTRUM

For the group working under Richter the discovery of charmonium (J/ψ) was only the beginning. Eleven days later they capped off the discovery with a second. As in the case of the first particle they found a slight bump at 3.7 GeV, then narrowing in on it as they had previously they found it was also a high sharp peak. It was an excited state of the J/ψ, and was named ψ′ by the SPEAR group.

Once theorists got into the act the future for more such resonances looked bright. Not only were there likely other excited states of charmonium, but there was also a state in which the spins of the two quarks were opposite to one another (in J/ψ they are in the same direction).

Then came another breakthrough. David Politzer, one of the discoverers of asymptotic freedom, teamed up with Thomas Appelquist of Harvard University and the two noticed that charmonium was much like another well-known combination—positronium. Positronium—a bound state of an electron–positron pair—had been observed 25 years earlier. And over the years it had been thoroughly studied. Much was now known about it. It was so similar to the hydrogen atom—just a hydrogen atom with the proton replaced by a positron—that the theory of the hydrogen atom could be used in analyzing the new model. The energy-level spectrum of the positronium model had been worked out in considerable detail. Similar diagrams were soon drawn for charmonium. They were referred to as the charmonium spectrum.

Experimentalists now had a guide—something concrete to look for. But their search proved to be much more difficult

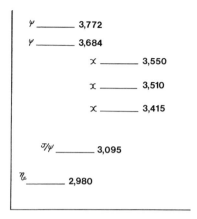

The charmonium spectrum. Numbers to the right are energies of the levels.

than expected. Months passed with no results. Besides SPEAR, the German electron–positron annihilator called DORIS was pressed into the search. Both J/ψ and ψ' were quickly found by the DORIS team, but when they turned to the newly predicted states they found nothing. Six months passed and still nothing. Then finally a breakthrough came from DORIS. Theorists had shown that the ψ' should decay to an intermediate state before it finally decayed to light particles. Experimenters at DORIS found the state; shortly thereafter the SPEAR group also found it. It was a small success but it did much to bolster their sagging egos. Their elation, however, was short-lived. Theorists at Johns Hopkins and at Imperial College in London showed that everything that had been discovered so far could be explained without introducing charm. Most scientists believed that their arguments were contrived, but they had to be taken seriously.

There was, however, a foolproof way to verify charm, and that was to find "naked charm."

NAKED CHARM

As we just saw charmonium had hidden charm: the charm of the antiquark canceled that of the quark. With zero net charm the J/ψ would not show charmed behavior. But there were three other quarks: u, d, and s. And if charm did, indeed, exist it would have to couple with them. Combinations such as cū and cd̄ should exist, and if they did, they would have naked charm.

Calculations indicated that the lowest energy naked charm state should be somewhere near 2 GeV. A search for it was initiated at SPEAR. But again months passed and nothing was found. Experimentalists began to get frustrated. "Do the theorists really know what they are talking about?" They moved up the energy spectrum to 4 GeV and continued the search, then to 7 GeV. Still nothing.

Gerson Goldhaber of SLAC had had enough. What was needed, he decided, was a new approach. Why, he asked himself, weren't they seeing the postulated particles? He soon convinced himself that it was the design of the detectors. It was quite possible that they were unable to "see" them. Looking again at the type of reactions to be expected he decided to go back through the records and look for them carefully. About the same time a French physicist, Francois Pierre, initiated an independent search. Within a short time both men had found examples of reactions that should have involved charm. When Goldhaber heard about Pierre's success he joined forces with him and the two of them found dozens of similar events. Others then joined in the search. Soon there was no doubt: charm had been found. It had, in a sense, sneaked in the back door, nevertheless it had been found.

What Goldhaber and Pierre had found was evidence of combinations such as cū and cd̄. Such states are referred to by the letter D. A paper reporting the existence of the D mesons was submitted for publication in June 1976.

Theorists were delighted with the discovery of charm and its associated spectrum not only because it verified their theo-

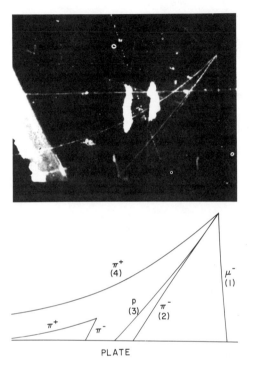

An example of the production of a charmed baryon. The charmed particle is not seen directly, but its decay products, p, π^+, π^-, are visible.

ries, but because they expected it would also explain the rise in the ratio R. R would, in essence, be brought in line with theory. But they were incorrect. R was still slightly too high, which concerned them. The rise in R could indicate the existence of another quark, one that had not yet been discovered, or perhaps another lepton, a heavier one.

Theorists didn't want another particle because the whole idea of introducing charm was to give a symmetry to the quark–lepton scheme. At the time there were four leptons and three quarks. The fourth quark, charm, brought this up to four each—

THE TAU LEPTON

In 1974 Martin Perl, one of the group leaders at SPEAR, set out to see if he could find a lepton that was more massive than the muon. His first problem was how to detect it. Realizing that if it had a mass greater than about 1 GeV it would have to be detected indirectly because it would decay before it reached the detector, he considered the most likely type of event that would produce it. If heavy leptons were produced in an electron–positron collision, he decided, the most likely decay products would be electrons and muons. Neutrinos would also be produced but he could disregard them as they wouldn't be seen. He set out, therefore, to look for "electron–muon" events. And by late 1974 he had detected several. Delighted with the discovery, he informed many of his colleagues at SPEAR. But they weren't convinced. To them, all he was seeing were electrons and muons, and there were several other ways they could be produced. One of them, in fact, pointed out that the D meson, discovered only a few months earlier, would decay to electrons and muons. Undaunted, Perl was confident his evidence indicated a new heavy lepton—a tau (τ) particle, as he called it. And in 1975 he announced the discovery.

Still, few were convinced. So Perl continued to look for more events, and in 1976 he was successful. But then came a serious setback that created even more skepticism. A group at DORIS, with equipment at least equal to that he was using, reported that they had looked for heavy leptons and had found none.

It was a serious setback for Perl and his group, but they continued on with their search. Then came the first ray of light.

One of the other groups at SPEAR discovered a large number of muon–hadron events that strongly indicated the existence of a tau. Then the following year experimenters at DORIS reported that they had seen both muon–electron and muon–hadron events. There was now little doubt: the tau particle, a heavy lepton, had been discovered.

When a check was made of the ratio R, it was found that the tau completely accounted for the discrepancy (the fact that it was too high). Thus, one crisis was over—but it was quite obvious that the solution had created another. Associated with the tau was a neutrino, bringing the total number of leptons to six. And there were only four quarks. The symmetry was gone!

UPSILON

Another quark, in fact, two more quarks were needed to bring the required symmetry back into the scheme. The obvious place to look for them was in electron–positron collisions. And nobody had to convince experimenters at SPEAR and DORIS that a search was needed. As soon as the discovery of the tau was announced they swung into action.

Concentrating on the area above 5 GeV they swept carefully through the spectrum up to the limit of their instruments, which was about 8 GeV. And they found nothing. There were no narrow peaks, and no indications that they were missing one. Furthermore, there was no longer a problem with R as there had been before the discovery of tau. Within the energy range of both SPEAR and DORIS it was almost exactly in line with theory.

Did this mean there was not another quark? If so, how would theorists explain the lack of symmetry? There was, of course, the possibility that the new particle lay above the energy range of SPEAR and DORIS. There was no electron–positron annihilator available for this energy range, but there was an

Leon Lederman, presently director of Fermilab. (Courtesy Fermilab.)

instrument that could produce such pairs in high-energy proton collisions (as had been done in Ting's experiment). Fermilab's giant accelerator was capable of producing 500 GeV.

With energy to spare, a group under Leon Lederman (now the director of Fermilab) set out to find a heavier quark. Lederman decided to observe not only electron pairs but also muon pairs. Large numbers of such pairs would also indicate a heavier quark. In late 1975 they observed a sudden increase in electron pairs around 6 GeV. But SPEAR and DORIS had already searched this region unsuccessfully. Had they discovered something SPEAR and DORIS had missed? Lederman was skeptical and realized that many checks would be needed. Nevertheless, he named the particle upsilon (Y).

In 1977 changes were made in the accelerator and Lederman

and his group again looked for the 6 GeV peak. It had disappeared. It wasn't a real peak after all. He continued scanning up the range and at 9.5 GeV he discovered another peak—this one associated with muon pairs. This was above the range of SPEAR and DORIS. Checks were made and it was soon obvious that this was not a quirk of the instrument—it was a real peak. Lederman then switched the name he had used earlier to this particle and announced the discovery in August 1977.

Like charmonium this particle was also a combination of a quark and an antiquark. The new quark was called bottom; this meant upsilon was a combination of a bottom quark and an antibottom quark (b$\bar{\text{b}}$). With bottom the number of quarks was brought up to five—still one less than the number of leptons. Was it possible that there was another quark? If we look back at the other quarks we see that they seem to come in pairs; up and down are associated with one another, as are strange and charm. It seems reasonable on the basis of this that there is another quark associated with bottom. It has, in fact, been given a name (the most obvious one, perhaps): top. So far, though, top has eluded experimenters. But it is believed that it will eventually be found.

SUMMARY

With the discovery of several new particles the theory became much more complex. Yet there was a certain beauty about it. There were three pairs of quarks, which we usually denote as

$$\begin{pmatrix} u \\ d \end{pmatrix} \quad \begin{pmatrix} c \\ s \end{pmatrix} \quad \begin{pmatrix} t \\ b \end{pmatrix}$$

And there were three pairs of leptons,

$$\begin{pmatrix} \nu_e \\ e \end{pmatrix} \quad \begin{pmatrix} \nu_\mu \\ \mu \end{pmatrix} \quad \begin{pmatrix} \nu_\tau \\ \tau \end{pmatrix}$$

We refer to $\binom{u}{d}$ and $\binom{\nu_e}{e}$ as first-generation particles, and $\binom{c}{s}$ and $\binom{\nu_\mu}{\mu}$ as second-generation, and $\binom{t}{b}$ and $\binom{\nu_\tau}{\tau}$ as third-generation. The well-known particles of our world are all first-generation. The quarks of each generation come in three colors, and aside from mass they are similar. The muon (μ) and the tau (τ) are also identical to the electron except for mass. They are essentially "heavy electrons."

A question that immediately comes to mind is: How many more quarks and leptons are there? We still do not know but there are indications from cosmology that no more than one more pair in each group would be possible. Another obvious question: Is there any connection between the two families? In other words, are quarks and leptons related? We'll leave that to a later chapter.

CHAPTER 10

Search for the W

During the mid 1970s SPEAR and DORIS led the way in discoveries. Both were electron–positron annihilators. Then in 1978 a new and larger machine called PETRA was built to replace DORIS. It was capable of producing energies up to 38 GeV. A couple of years later SPEAR at Stanford was superseded by an instrument called PEP. It was capable of generating energies up to 36 GeV.

As these new, higher-energy machines came into existence physicists began to wonder: What should they look for? One of the most obvious things was, of course, the top quark. The bottom quark had been discovered a few years earlier and physicists were sure that another quark of slightly higher energy was associated with it. But after an extensive search it was not found. In fact, even when higher-energy machines came into existence later it still was not found. Another possibility was the exchange particle of the weak interactions, the W particle. But calculations indicated it was likely out of range of either instrument.

Interestingly, what was discovered was something entirely unexpected, a phenomenon we now refer to as "jets."

JETS

What exactly is a jet? To answer this let's begin by examining what happens when an electron collides with a positron.

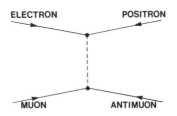

An electron–positron collision creating a muon–antimuon pair.

The two particles will, of course, annihilate one another, creating in their place a virtual photon. If the energy is not too high, of the order of, say, a few GeV, the virtual photon may create another electron pair, or it may produce a muon pair. At higher energies we will get quark pairs. And this is where jets come in. Upon being created the quark and antiquark will begin to separate. But wait! I mentioned earlier that quarks are held together by strings, and it was virtually impossible to break these strings. How, then, are the quark and antiquark able to separate? The best way to answer this is to look at a particular case, say the generation of charm–anticharm pairs ($c\bar{c}$). We'll assume that the photon is energetic enough to both produce the charmed pair and give them energy to separate. As they move apart the string holding them together will stretch. In an earlier chapter I said that the string then breaks and a quark and an antiquark appear at the end points. This is one way of looking at what happens, but there is a better way. We saw earlier that there are many virtual particles in the vacuum: electrons, positrons, muons, and quarks. Energy is, of course, needed to make them real, but the escaping quark and antiquark have energy. The c quark can therefore "pick up" an antiquark out of the vacuum. Let's assume it is a $\bar{u}$. It then becomes $c\bar{u}$, which we know is a D meson. And once it becomes a meson the string, of course ceases to exist. At the same time, the antiquark ($\bar{c}$) picks up a quark, say a d quark, and becomes $\bar{c}d$, which we know is also a D meson. What we should see, then, is the emission of two D mesons.

And in practice this is what we do see. In fact, we usually see many more than just two D mesons. Pions, kaons, and even heavier particles are also frequently observed. In some cases a dozen or more particles come out, all generally in the same direction. This is the "jet" I referred to earlier.

Actually, because of conservation of energy and momentum, we get two jets, back to back. We sometimes refer to these jets as hadron jets, since they are made up of hadrons. Furthermore, since the hadrons themselves are made up of quarks, we are, in a sense, seeing the quarks themselves.

The first evidence that such jets existed came from SPEAR in 1975, just after the discovery of the J/ψ particle. But there was great uncertainty as to what was actually being observed, and few were convinced that they were seeing jets. There was no question, however, after PETRA became operational. Jets existed.

For a while the so-called two-jet events caused considerable excitement. But then three CERN theorists, John Ellis, Mary Gaillard, and Graham Ross, predicted that three-jet events should also exist. They showed that it was possible that a gluon, emitted by either the quark or the antiquark of the pair, could create another jet. The hunt was then on for three-jet events. It

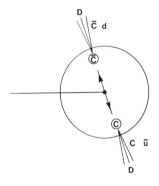

Jets created when a charm–anticharm pair separate generating D mesons.

was three years, though, before they were found. And if we look at the details of the experiment it is not hard to see why. In the case of two-jet events the jets come off back to back, and their directions can be predicted. In the case of three-jet events the particles are emitted in arbitrary directions that cannot be accurately predicted. This means that many detectors were needed to observe them.

Despite the difficulties, experimentalists at PETRA came through, and in mid 1979 the first three-jet events were seen. Theorists were jubilant, for not only was it a verification of their prediction, but it was also a verification of quantum chromodynamics, as Gaillard, Ellis, and Ross had used QCD in making the prediction.

Most of the important experiments during this period were done using electron–positron annihilators, but they were not the only type of accelerators in operation. In 1971 a large proton collider called ISR (Intersecting Storage Rings) at CERN went into operation. And at Fermilab a 500-GeV proton accelerator went into operation in 1972. As we will see, though, proton–proton collisions are much more difficult to deal with than electron–positron collisions.

Electron–proton collisions are simple because the electron is considered to be a point particle. It has no observable substructure and is therefore much less complicated than the proton. The proton, as we saw earlier, is made up of three quarks, which have gluons moving back and forth between them. Let's begin, then, by assuming that the incoming electron has enough energy to knock one of the quarks out of the nucleus. What happens? The quark is, of course, attached to the other quarks by a string, but because of the energy imparted to it by the collision it is able to pick up an antiquark in its flight out of the nucleus. And once it becomes a meson the string disappears. In practice, as in the previous case, many mesons are generated and we observe a jet. There are, of course, two quarks left back in the nucleus, but since there are no stable two-quark particles, they soon decay and form new particles.

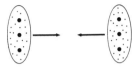

The collision of two protons. Large dots are quarks; small ones, gluons.

Now to the proton–proton collision. In this case we have a "bag" of three quarks and a number of gluons, striking a similar "bag." Several things can happen. Two of the quarks, one from each proton, for example, could strike one another head-on, with the other quarks missing one another. In this case we would expect debris in the two directions along the beam (from the quarks that passed through generally undisturbed). But the quarks that strike one another will produce jets that come off in a direction perpendicular to this. Therefore, we should see two jets perpendicular to the beam, and two along the beam's direction. Such events have been seen at both ISR and Fermilab.

Things can, of course, be much more complicated than this. It is possible that two of the quarks in one of the protons could strike two in the other. In this case we would expect four jets perpendicular to the main beam.

And there are several other possibilities. Quarks from one of the protons could strike gluons in the other, and also gluons in one could strike gluons in the other. Needless to say, situations like this are a theorist's nightmare.

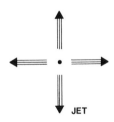

Production of jets in a proton–proton collision.

SEARCH FOR THE W

With the introduction of even higher-energy accelerators at Fermilab and CERN (SPS) in the early 1970s attention turned to the W particles: W^+, W^-, and Z^0. We talked about them earlier; they are the exchange particles of the weak interactions. As you may remember the W is generated in weak interactions when there is an exchange of charge; neutrinos change into electrons in this case. When there is no charge transfer we are dealing with the neutral current case, and the exchange particle is the Z^0. We saw in the last chapter that neutral currents were shown to exist in 1973.

The next step was obviously a verification of the existence of the particles ($W^\pm$, Z^0) themselves. But even with Fermilab's giant accelerator and CERN's SPS (Super Proton Synchrotron) they were still beyond our reach. This seems a little strange when we look at the details. Calculations showed that the W and Z particles had a mass in the range 50–100 GeV, and we should therefore be able to generate them if we have this much energy. Fermilab's accelerator was capable of producing 500 GeV and the SPS at CERN was capable of 400 GeV. This is well beyond the 50–100 needed.

Why, then, were scientists unable to see them? To answer this we have to look at what happens to the energy of the particles after the collision takes place. At the time both of the above instruments were proton accelerators. This meant that the accelerated protons hit other protons in a fixed target. In such collisions most of the energy goes into the kinetic (moving) energy of the debris particles, and very little is left over to produce new particles. It turns out that only about 28 GeV is actually available to produce new particles—and this is far below the 50–100 GeV needed.

The only machine that would be capable of producing such energies was one that was planned for CERN called LEP (large Electron Positron Collider). It was to be an electron–positron collider with an available energy of 100 GeV, but it would not be ready until the late 1980s.

Carlo Rubbia.

Carlo Rubbia of CERN (and Harvard) was not interested in waiting. And neither were many Americans at Fermilab. Each knew that if they delayed too long the other would have the honor of discovering the W. In 1976 Rubbia along with David Cline of the University of Wisconsin and Peter McIntyre of Harvard began looking into the possibility of a shortcut. The large proton accelerator SPS had just become operational, but it was incapable of providing the required energies. Would it be possible to accelerate antiprotons, in addition to protons, in the same (SPS) ring? If so, they would have a proton–antiproton collider at almost no additional cost. The protons would travel in one direction around the ring, the antiprotons in the opposite direction. And when the protons and antiprotons collided considerable energy would be available for producing new particles.

The SPS accelerator could accelerate protons to 270 GeV. If it could be used to accelerate antiprotons in the opposite direction, they would also be accelerated to 270 GeV. If the two beams were then brought together the total energy would be 540 GeV—well above the 50–100 needed.

Of course there was still the problem that some of the energy would end up as kinetic energy. But calculations showed that the amount available for the generation of new particles was easily enough to produce the W.

With the first hurdle out of the way Rubbia and his group turned to the second: the generation of an adequate beam of antiprotons. The production of the antiprotons themselves was no problem. They could be generated by merely slamming protons into a metal target. In practice, several types of particles would be generated, but the antiprotons could be easily separated out using magnetic fields.

There was, however, a serious problem once they were separated. The antiprotons were produced with many different velocities, and were therefore traveling in many different directions. If they were injected into an accelerator like this most of them would soon slam into the sides of the chamber and be lost. The beam obviously had to be streamlined before its energy could be increased.

Fortunately, a technique for doing this was available. In fact two techniques were available. The one that proved best for Rubbia's group was called "stochastic cooling." In this method the beam was "cooled" (streamlined) by applying small "kicks" to it. In practice a "bunch" of antiprotons would be injected into a ring where a sophisticated control system would monitor it, looking for deviations from an ideal orbit. When it detected them it would send a signal to a control that would make the appropriate corrections.

Meanwhile, similar experiments were going on at Fermilab in the United States. Experimenters were also working to convert the proton accelerator into a proton–antiproton collider. But their facilities differed from those of CERN's and they were not able to use the same cooling technique. They had to use a more complicated one. Because of this, and financial problems, their project soon began to lag behind.

Soon after proving that they were able to produce a satisfactory pulse of antiparticles the CERN team developed a technique for accumulating them. Once a pulse, or "bunch," of antiprotons was cooled it was put in a holding orbit in the ring, or accumulator as it was called. Another bunch would then be put into the accumulator and cooled; this bunch would then be added to the first bunch. And so on. Eventually a large accumulation would be built up. After about two days, for example, the orbiting bunch consisted of about 60,000 small bunches.

At this stage the bunch had an energy of only a few GeV. It was then sent to an accelerator called PS (Proton Synchrotron) that increased its energy to about 26 GeV. Protons were also introduced into PS (in the opposite direction) and were accelerated to 26 GeV. Finally both the protons and the antiprotons were fed into the larger SPS ring where they were taken to 270 GeV. With that they were ready for the first experiment.

By now Fermilab was no longer in the race. They had had problems with cooling, and had eventually decided to use a new type of magnet in their accelerator, a much more efficient one called a superconducting magnet. In the long run this was an important development, but it took them out of the race for the W. They would eventually have more energy than CERN but the superconducting ring was not complete until the spring of 1983.

With the production of the first proton–antiproton collision on July 9, 1981, experimenters at CERN were ready to search for the W. They soon found, though, that another roadblock lay in their path. The antiproton bunches did not contain enough particles to produce an adequate number of W's. When the protons and antiprotons came together only a few would interact—most would just continue on their way around the ring. And with the number that were colliding they could expect to see only about one W a year. They obviously had to increase the density, or "luminosity," as it was called, of the antiproton beam.

Before we continue let's look at how they expected to detect the W's. The W's themselves would be generated in the proton–antiproton collision but they would not live long enough to be

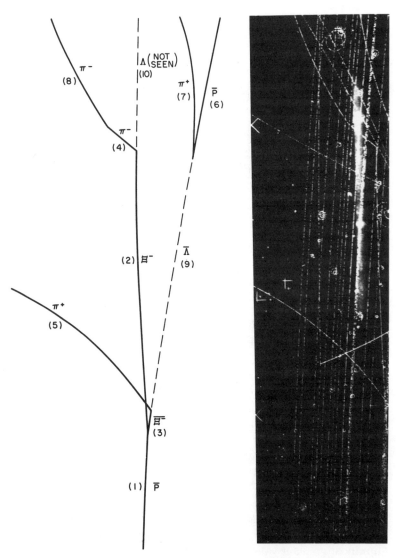

Tracks showing a proton–antiproton collision. The antiproton ($\bar{p}$) is incoming at the bottom. It collides with a proton at (3). (Courtesy Brookhaven.)

seen directly. Their presence would have to be inferred from their decay products. This was, of course, nothing new. Many particles are detected in this way. Calculations showed that the W could decay to a positive muon and a neutrino. Or it could decay to a negative muon and an antineutrino. Another possibility was the production of electrons (positrons) and antineutrinos (neutrinos). The problem, then, was to build a detector that would detect these particles. And late in 1977 work began on one. It was a huge project that eventually involved over 100 scientists and engineers. Scientists from nine countries, including the United States, worked on the project. The detector was called UA1 (underground area one)—a reference to its position underground.

A huge underground cavern had to be excavated to house the detector. The detector itself was designed to fit around the beam pipe. It could be rolled up to the pipe on rails when needed, and stored away in an underground garage when not in use.

There was so much excitement about UA1 and the W experiment that a second detector called UA2 (underground area two) was assembled. It was similar to UA1, but not as sophisticated. The group building it, however, was equally anxious to detect the W and Z, and features were designed into it with this in mind. It was different from UA1 in that it had no magnetic field in its central section and therefore particle tracks were straight in this region, rather than curved as they would be in a magnetic field.

SUCCESS AT LAST

When the first experiments began in July 1981 the luminosity of the antiproton beam was not high enough to see W's, so scientists went to work trying to improve it. Six months later they had increased it by a factor of about 50, but this was still too low.

The SPS accelerator was then shut down. Rubbia and his colleagues began to get worried. The proton–antiproton annihilator at Fermilab was still a long ways off but it was possible that they could be beaten. In the spring of 1982 SPS was turned on again. Then came a series of serious setbacks. First UA1 was contaminated with dirty air and had to be disassembled and thoroughly cleaned. Then a large waterpipe burst and flooded the underground cavern to a depth of about six feet.

It was not until October 1982 that things got back to normal. Frustrated by the delay the group worked fast now. The luminosity was soon increased by a factor of 100. The detector would now be able to detect a W every ten days. Data were taken and each of the teams began analysis as fast as they could. It wasn't an easy task, but they were helped by the computer. A trigger in the detector had been set up to record any event that looked reasonable. And by December 1982 UA1 had 140,000 of these events. They were fed directly into the computer where a series of tests were applied.

The computer had soon narrowed the 140,000 down to less than 100. Finally, there were only 34 left. Each of these was then viewed directly, and 29 of them were discarded. This left only 5, but they were gold: they had all the criteria required of a W. Four of them showed the decay products of W^-, one showed those of a W^+. The UA1 team was jubilant. But there was a final important check. Was the energy of the event in the proper energy range? By determining the energy of the particles in and out of the reaction they were able to squeeze in on it. The result: 81 ± 5 GeV, in almost exact agreement with the Weinberg–Salam theory. They had found the W!

Meanwhile the UA2 team was analyzing their results using a similar computer program. Narrowing the events down in the same way they were eventually left with four events. Further evidence for the W. Their energy was also approximately 80 GeV.

On January 25, 1983, the news was announced. The W had been found—or at least strong evidence for its existence had been found. But there was still the Z^0. Could it also be detected using the same equipment? The collider was now closed down for maintenance so they would have to wait until spring. In the case of the Z^0 they would have to look for electron and muon pairs. And as soon as the accelerator came back on in the spring they began their search. On May 4 they found the first interesting event. Then later in the month five more were detected. In June the announcement was made: the Z^0 had also been found.

THE HIGGS PARTICLE

As we just saw, one of the products of the proton–antiproton collision is the W (and Z), but it is not the only one. According to the Weinberg–Salam theory when the W and Z gain weight through what is called the Higgs mechanism a Higgs particle is left over. Would this Higgs particle show up in the experiment? A search was made for them but they were not found. The major problem with the Higgs is that we do not know its mass, so we have no way of knowing if it is even in the range of the proton–antiproton collider. Furthermore, although it is an integral part of the Weinberg–Salam theory there are physicists—most notably M. Veltman—who are not convinced they exist.

Another problem is that if the Higgs has a mass greater than 1000 GeV the W's and Z could not be considered to be elementary particles, but rather would be composites. If this turns out to be the case it would throw a wrench into the theory. We will not know if it is the case until larger accelerators are built.

Anyway, with the discovery of the W and Z an embarrassing hole in electroweak theory had been plugged and theorists could push forward in their efforts to unify electroweak theory with QCD, the theory of strong interactions. We will turn to this in the next chapter.

CHAPTER 11

Unifying

Theorist and experimentalist alike could lean back in their armchairs and gloat. The electromagnetic and weak theories had been unified, and a theory of the strong interactions (QCD) had been devised. Taken together, these two theories are usually referred to as the "Standard Model."

Within this Standard Model were two families of particles: the quarks and the leptons. And if we neglect gravity for the moment there were also only two forces within it: the electroweak and strong nuclear. But was it possible that we could go a step further? Could we combine the strong nuclear and electroweak forces? Indeed, could we also combine, or unify, the two particle families? If so, we would have a family with one particle in it and one force. Such a family is the dream of particle physicists, for it would be a unified theory of nature.

But there were other reasons for wanting to unify QCD and electroweak theory. Taking a close look at them we see that, although they are beautiful, well-constructed theories, they do contain defects. The electroweak theory, for example, is not truly unified in that in relies on a "mixing parameter" to determine how the electromagnetic and weak fields are combined. And this is not the only arbitrary parameter within the theory.

Also unexplained is the quantization of electric charge. And there are further problems: Why are there charges that are opposite one another, and why do quarks have a charge one-third that of the electron? These were important questions that theorists believed might be answered with a unification of the two theories.

Before jumping into the details of such a unification, though, we should ask ourselves what we would require of it. First, it would have to be a gauge theory. Both QCD and electroweak theory are gauge theories and they work exceedingly well. Furthermore, if these theories are to be brought together we are going to need a group that contains the groups they are based on. Electroweak is $SU(2) \times U(1)$ and QCD, $SU(3)$, so our new group would have to contain these as subgroups.

The first to attempt such a unification were Jogesh Pati of the University of Maryland and Abdus Salam of Imperial College in late 1973. Pati and Salam had been longtime friends, having worked off and on with one another ever since they first met in the late 1950s. Their first job, they decided, was to unify the two families of particles—in short, to bring them together into one family. This meant that leptons were just quarks in disguise. But how could they be brought together? Clearly, one of the simplest ways was to just let leptons be quarks of a different color (i.e., assume they are actually the same type of particle). And this is exactly what they did. Instead of three quark colors they had four. Using this as a starting point they developed a theory based on the group $SU(4) \times SU(4)$.

Before long, though, it became obvious that there was a problem: if the quarks and leptons were in the same family they had to be able to change into one another. In other words, under the proper conditions, a quark should be able to change into a lepton, or a lepton into a quark. The major difficulty, though, were the consequences of this when it was applied to the proton. Inside protons are three quarks, and if one of them changes into a lepton the "proton" is no longer a proton. In short, what this implied was that protons, like all heavier particles, decayed.

It was obvious to Pati and Salam, though, that the lifetime of the proton had to be extremely long. After all, the Earth had

had been around about 10^{10} years and it wasn't showing any signs of decay. They didn't like this aspect of the theory but realized that they couldn't neglect it if they were to publish. "We had violent arguments about what to do," said Salam. But after realizing that the lifetime had to be greater than about 10^{28} years Salam said he felt more confident.

The two men therefore wrote up the theory and submitted it to *Physical Review Letters*. Although it had many loose ends the overall idea was new and interesting. Salam was sure it would be accepted.

But back it came—rejected. It wasn't that the editors felt the paper was poorly constructed or incomplete—it was just inappropriate. *Physical Review Letters* was originally set up because of the long publication delay associated with *Physical Review*—sometimes as long as a year. Important discoveries could be announced in the *Letters*, but there had to be an urgency associated with them. The editors didn't feel Pati and Salam's article was urgent.

Salam was outraged. He promptly went over the head of the editors and managed to get it published. Still, even when it appeared, it was hardly noticed. Few were interested. And even fewer believed that the proton could decay.

The lack of interest did not deter Salam. He continued to push the idea wherever he went. But, perhaps because he was also pushing a few of his other pet ideas at the same time, no one took him too seriously. He disliked Gell-Mann's fractional charges and refused to incorporate them into his theory. Thus, the Pati–Salam theory contained quarks with integral charges. Furthermore, he hated confinement even more, and preached "quark liberation" at every conference he attended.

But "unification" was not an idea that was going to die easy. The following year (1979) Sheldon Glashow teamed up with Howard Georgi, a postgraduate student at Harvard, and together they tackled the same problem. Like Pati and Salam they also realized they would have to find a larger group that would contain $SU(2) \times U(1)$ and $SU(3)$ as subgroups. But Georgi was an excellent group modeler. (He is the author of one of the best-known books on the application of groups to particle physics.) They began by listing all the requirements of a unified theory. Both were familiar with the Pati–Salam paper and felt that it was well constructed—but seriously flawed. The large difference in strength between the three force fields had generally been ignored. More particularly, the corresponding differences in the coupling constants had been ignored. And Glashow knew that this would have to be explained.

After deciding what they required of a unified theory Georgi and Glashow went to work trying to construct one. But as Georgi said, "We couldn't agree on anything." Everything that Glashow came up with Georgi shot full of holes, and everything Georgi suggested Glashow disliked. When they finally separated after a trying day both men continued working far into the evening on their own. Georgi tried several groups that looked promising, but none worked. Then he tried the five-dimensional group $SU(5)$. It had the first requirement: it contained $SU(3)$ and $SU(2) \times U(1)$ as subgroups. Pleased, he began to look into it further, and the more he looked the more pleased he became.

Because it was five-dimensional he needed five basic particles with which to build all other particles (within the representation). After a little juggling he found that the three colored down quarks, the positron, and the antineutrino worked, assuming they spun in the same direction. He could account for all possible interactions between them by using the known exchange particles: the photon, W, Z, and the gluons. The W particles converted electrons (positrons) into neutrinos (antineutrinos) and vice versa; the gluons converted one color of quark into another. But he found he also had to introduce a new type of particle—a particle that would change quarks into leptons. He called it the X particle. There had to be twelve different varieties of these particles.

	d^R	d^G	d^B	e^+	$\bar{\nu}$
d^R	G, γ, Z	G	G	X	X
d^G	G	G, γ, Z	G	X	X
d^B	G	G	G, γ, Z	X	X
e^+	X	X	X	γ, Z	W
$\bar{\nu}$	X	X	X	W	Z

X, X particle
G, gluon
γ, photon
W, Z, exchange particles of weak interactions

Everything came together like a jigsaw puzzle. Georgi was delighted to find that all spaces were filled and there were no particles left over. In short, everything fit.

There are, of course, more than five elementary particles. Where do the others come in? Georgi showed that he could generate ten more using the above five. And they had exactly the properties of ten known particles. Pressing on, he found he could get all the particles associated with the first generation (i.e., $\binom{u}{d}$, $\binom{\nu_e}{e}$). Then, applying the same procedure he went to the second generation particles (i.e., $\binom{c}{s}$, $\binom{\nu_\mu}{\mu}$) and showed that their associated particles could also be accounted for. Similarly for the third-generation.

There was, however, a problem (the same one Pati and Salam encountered): the proton decayed and there was no way he could get around it. The next day when Georgi got together with Glashow he explained his ideas. Glashow was enthusiastic until Georgi got to the decay of the proton. They were both sure, though, that this would not be an insurmountable barrier. Glashow had heard that Fredrick Reines of Los Alamos had just published a lower limit on the lifetime of the proton so the two men quickly looked up his paper. Reines had arrived at a figure of 10^{27} years. They were relieved to see that it was so long. But there was another worry: they were unable to calculate this number from the theory. (I should mention that Pati and Salam were also unable to calculate it.)

Glashow realized that if the proton's lifetime was indeed as long as Reines said, the X particle that caused the decay would have to be heavy—a thousand trillion times heavier than anything that had ever been discovered. It would be impossible to generate such a particle using modern-day accelerators. An accelerator as large as the solar system would be needed. Glashow chuckled to himself as he contemplated the announcement of such a particle. It would certainly stand most experimentalists on their heads.

The two men wrote up the paper and in June 1974 submitted it to *Physical Review Letters*. It appeared a month later. Most scientists, upon examining it, agreed it was a masterpiece of ingenuity that had been carefully and skillfully constructed. But when they got to the end of the paper and saw the prediction of proton decay their enthusiasm waned.

Despite its grand structure, the theory was incomplete; it didn't explain why the coupling constant of the strong and electroweak fields were different, nor predict the lifetime of the proton—but Georgi wasn't finished. Shortly after its publication he got together with Steven Weinberg and a post-graduate student, Helen Quinn, and they took a second look at it. Using the theory Politzer had used to predict asymptotic freedom, they were able to reconcile the coupling constants and come up with an estimate of the proton's lifetime. The breakthrough that allowed them to do this was the realization that the coupling constants were not actually constants. They varied, depending on the energy. The strong nuclear coupling constant, for example, decreased considerably as you went to higher energies.

What would cause such a strange behavior? The best way to answer this is to look at the virtual cloud around the particle. How is it affected by an increase in energy? The Uncertainty Principle tells us that there is a relationship between energy (or mass) and distance: the higher the energy (or the more massive the particle) the shorter the distance a virtual particle can travel. This means that associated with any energy there is an equivalent distance. For example, if we wanted to examine a particle of size 10^{-16} centimeter, we would have to have an energy of 100 GeV. Consider the effect of this on, say, the quark cloud. As you may remember, because of asymptotic freedom, when you increase the energy you spread out the quark's color charge. This means that at close range (small distances) its coupling constant will be smaller. In other words, if you go to higher energies the coupling constant will decrease. Exactly the same thing happens, only to a lesser degree, in the case of the weak interactions. In the electromagnetic interactions, on the other hand, we have the opposite effect: as you go to higher energies the coupling constant increases.

Georgi, Weinberg, and Quinn followed these changes with increasing energy and found that eventually all three coupling constants came together. This occurred at an energy of 10^{15} GeV. (The equivalent distance is 10^{-29} centimeter.)

With this we finally had an explanation of why the coupling constants are different. But how does this give us a unification? It is impossible for us to generate such energies, and likely always will be. However, if we go back to the first fraction of a second after the universe was generated in the "big bang" explosion that presumably occurred 18 billion years ago, we find energies this high. This means that at this time the coupling constants were all equal and all three forces of nature were the same. They were indistinguishable, and therefore unified. Unification, then, is not apparent in the universe of today, but it did occur when the universe was very young—less than a second old.

Incidentally, the unification energy, 10^{15} GeV, is also the energy that is needed to generate X particles. So it's obvious we're never going to generate them. Of course, if we could detect proton decay we would be detecting it indirectly.

You might ask: how did the X particle get such a large energy? Of the exchange particles, only the W's are massive.

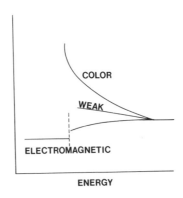

A simple representation showing the color, weak, and electromagnetic field strengths coming together at high energies.

And they get their mass from spontaneous symmetry breaking; they "eat" Higgs particles. Do W particles get their mass in the same way? Indeed, they do. But they have to eat a supermassive Higgs particle.

One of the predictions of grand unification [SU(5)] theory seemed, at first, to be in conflict with experiment. In electroweak theory we have a parameter called the mixing angle; it gives the mixing ratio of the weak and electromagnetic fields. There is no way we can determine this number from electroweak theory, but we can determine it experimentally and this has been done. Shortly after SU(5) theory was devised, though, it was noticed that this angle could be calculated. Theorists made the calculations and to their dismay it didn't agree with the experimental value. After the Georgi, Weinberg, Quinn paper appeared, however, theorists realized that they were calculating the number in the wrong way. Because the coupling constants vary, this number is different at different energies, and it wasn't being determined for an energy corresponding to our present universe. When the calculations were redone the numbers (experimental and theoretical) were much closer. Still, they were not exactly the same.

SU(5) also solved another problem that had plagued theorists—particularly cosmologists—for years: the asymmetry between matter and antimatter in the universe. It is difficult to determine how much antimatter there presently is in our universe, but there seems to be no doubt that matter predominates. Yet there is evidence that at the very early stages it had to have had equal amounts of matter and antimatter. SU(5) allows both of these situations to exist. The Russian physicist Andrei Sakharov was the first to realize this. He showed in that if the quantum number B (the baryon number) is not conserved, a matter–antimatter asymmetry could arise. And, of course, if quarks changed into leptons B cannot be conserved.

But as the implications of the new theory began to sink in a problem, or perhaps I should just call it an unpleasant surprise, came to light. At relatively low energies, those our accelerators are capable of, there are hundreds of particles. It is an "interesting" region, explained fully by the standard model [SU(1)×SU(2)×SU(3)]. As we go to higher energies, though, we find (theoretically) that there are no new particles. The region above 100 GeV seemed to be "uninteresting"—all the way out to 10^{15} GeV where we see the appearance of the X particles. And, as I said earlier, we will never reach this region. This casts a gloom on experimental particle physics. If it is the case, there isn't much of a reason to build larger accelerators—the SSC, for example—because we won't see anything. The Higgs particle might be visible, but not much else.

If we did find an elementary particle in this region it would mean that SU(5) isn't valid. Any way you look at it, though, with the prediction of what Glashow calls a "desert" beyond 100 GeV, it may be difficult to get larger accelerators funded. Looking at it realistically, though, we shouldn't be too pessimistic, as we're still not sure SU(5) is correct, and indeed, as we'll see later there are indications it is not.

Furthermore, SU(5) isn't the last word. It is only one of several grand unified field theories. At the present time it seems to be the best, and most natural, but there are others. So if SU(5) is eventually shown to be incorrect all is not lost. Besides, as we'll see in a later chapter there are promising alternative approaches.

PROTON DECAY

Earlier we saw that if quarks change into leptons, the proton must decay. At first there was little interest in measuring the rate of decay. Salam had trouble getting experimentalists interested, as did others. But when a number of the predictions of SU(5) appeared to be correct, experimentalists began to take notice.

One of the first questions that comes to mind in relation to such a decay is: what does it decay to? We know the proton is composed of three quarks: u, u, and d. And if one of them is to change into a lepton we have to have the transfer of an X particle between them. The one emitting the X particle will change into a new particle as will the one absorbing it. One way this can happen is if one of the u quarks emits it and is absorbed by the d quark. In this process the u quark will change to a u, and the d will become a positron. This will leave $u\bar{u}$, which we know is the neutral meson. This means that we would see, upon decay, the emission of a pion (π^0) and a positron (e^+).

If you stop for a moment to think about this process you realize that there is something odd. Earlier I mentioned that the X particle is extremely massive, corresponding to an energy of 10^{15} GeV. This is a million trillion times heavier than the proton. How can a proton have a particle this heavy inside it? It seems strange, but it is possible because of the Uncertainty Principle. The X exists, according to this principle, only while it passes between the quarks. And the quarks have to be exceedingly close together for this to happen—about 10^{-29} centimeter. Compared to the overall size of the proton, this is infinitesimal. And while it travels this distance, it is of course virtual, so we can never measure it.

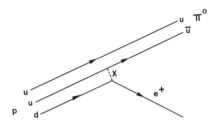

Proton decaying to a neutral meson and a positron.

Now, back to the proton itself. As we saw earlier Georgi, Weinberg, and Quinn predicted a lifetime of 10^{31} years. But the universe has only been around for about 10^{10} years. We obviously can't take a few protons and wait for one of them to decay. Surprising as it may seem, though, it is possible to do the experiment. If we had 10^{31} protons in a group we could expect one of them to decay each year. Better still, let's suppose we have about 10^{33}; we would then get about 100 per year, or one every three or four days. And, as it turns out, a group of 10^{33} protons isn't as large as you might expect—it's only about the size of a large house. Furthermore, protons are protons, and it doesn't matter what kind of material you use. You can select a relatively cheap material such as water or iron.

There is, however, another problem. If you select water, for example, and surround it with the appropriate detectors you will find you get an event every few seconds. Cosmic rays from space are continuously bombarding us, and they also react with the protons of the water. We would not be able to distinguish them from true decays of the proton. Therefore, we have to set up our experiment somewhere where it won't be bothered by cosmic rays. Since earth absorbs these rays, the obvious place is a cavern deep underground.

In hopes, then, of detecting the decay of the proton, several experiments were set up. One of the first was in India. A team of Indian and Japanese physicists used the Kolar gold mines in

India as a base of operations. Their detector consisted of 150 tons of iron located 12,000 feet underground. Within a year or so of setting up their experiment they reported that they had found an event consistent with proton decay. Considerable interest was generated. But then other experiments went into operation.

A European experiment, supported by CERN, was set up in a deep tunnel under Mount Blanc in the Alps. Because the whole mountain is above it, cosmic rays are easily excluded. In America most scientists looked to mines. A salt mine, about 2000 feet underground, was found under Lake Erie that was available. Water was used in this experiment.

Two other American experiments are now also in operation, one in Minnesota and one in Utah. So far none of these experiments have reported a decay in the range of the Kolar experiment. Indications are that the lifetime has to be greater than 10^{32} years—and this presents a problem. If it is any longer than this it may not be possible to detect it. Furthermore, it now seems as if it is longer than the prediction made by SU(5).

CHAPTER 12

Looking Deeper

We have seen how unification was used in an attempt to get around some of the difficulties of the standard model. Some of the problems were solved by these theories, but not all. In fact, the most promising of them, SU(5), now seems to be in serious trouble; it predicts a lifetime for the proton that appears to be too short.

I would like to reemphasize, though, that the standard model is an excellent theory. There are no conflicts between it and experiment. To be sure, there are predictions that have not yet been fully verified. The real difficulty, though, is not the "loose ends," but the theory itself. Within it are eighteen colored quarks and six leptons; furthermore, twelve exchange particles are needed to account for the interactions between them. We originally introduced quarks because the number of "elementary particles" was getting out of hand. And the three quarks that were introduced were certainly an improvement over the hundreds of "elementary particles" that were known at the time. But now it appears that things are getting out of hand again.

Besides the large number of particles, though, there are still a lot of unanswered questions. I discussed some of them in the last chapter. One of the most important is: why are there three generations of fundamental particles? We have $\binom{u}{d}$ and $\binom{\nu_e}{e}$ in the first generation, and all the well-known particles of our universe are made from them. But the second-generation particles,

$\begin{pmatrix} c \\ s \end{pmatrix}$ and $\begin{pmatrix} \nu_\mu \\ \mu \end{pmatrix}$, are exactly the same as them except for mass. The muon, for example, is nothing more than a heavy electron. And we have a similar situation for the third generation. How are these generations related? Why are there three of them?

Not only are there three generations, but everything seems to come in threes. There are three leptons with electric charge -1, three neutral leptons, three quarks of charge $+2/3$, and three quarks of charge $-1/3$. Also, there are three colors. Why so many threes? It is important that we be able to answer this.

Another difficulty is related to spontaneous symmetry breaking. Associated with it is the Higgs particle, a particle we have not yet observed. In fact, we can't even predict its mass. But, as we saw earlier, if its mass is much above about 100 GeV the W particle cannot be elementary; it has to be composite. Furthermore, there are a few theorists who are not convinced that it even exists.

TECHNICOLOR THEORY

The above difficulty with spontaneous symmetry breaking bothered Steven Weinberg now of the University of Texas and Leonard Susskind of Stanford University. Independently in 1979 they decided to see if there was any way of getting around it. And what they came up with was "technicolor theory."

The exchange particles of electroweak theory, the W's, get their mass by "eating" a Higgs particle, a particle that was invented strictly for this purpose. Technicolor theory is an attempt to give the W particle mass without resorting to a Higgs particle. The theory predicts a family of technicolor particles: techni-fermions, techni-gluons, and so on. These particles could not, of course, be visible at energies below about 100 GeV—the current limit of our accelerators. Otherwise we would have seen them. But they should appear when larger accelerators such as the SSC are built.

The W particle gets its mass in this new theory by absorbing "techni-mesons." But wait. Isn't this a bit odd: we have merely substituted one unknown particle (the techni-meson) for another (the Higgs). But mathematically this does get around symmetry breaking—and that's what we want. This doesn't mean, though, that the new theory is sound. It isn't. The major problem is the particles that carry the technicolor force—the techni-bosons. Their mass is generated in roughly the same way the W particle's mass is. Furthermore, calculations indicate that the techni-boson might be within the limit of present-day accelerators. But so far we haven't seen it.

So, while technicolor theory is an interesting alternative and has possibilities, it looks like a long shot at the present time.

PREONS

Technicolor theory was invented to overcome the problems associated with spontaneous symmetry breaking. But what about the other problems, the large number of quarks and so on, that we talked about earlier? Has there been any attempt to overcome them? Indeed, there has. Several theories have been advanced with them in mind.

Most of these theories take what might be considered the most straightforward approach: they assume there is a substructure underlying the standard model. What this boils down to is assuming that the quarks and leptons are made up of simpler particles. This approach certainly isn't new; it has been used since the dawn of science. We discovered the atom, then noticed that it had a substructure: the nucleus and electrons. Then we discovered the nucleus had a substructure: protons and neutrons. And finally we discovered that protons (and other hadrons) were made up of quarks. It seems logical, then, to apply this same approach to quarks.

But wait! This time things aren't quite as simple. Earlier I mentioned that leptons are point particles; in other words they have no structure. This also goes for quarks. And if they have no structure it doesn't make much sense to talk about the particles that they are made of. Fortunately, there's a way around this. What I actually meant was that they have no "observable" structure. And since the limit of observation is about 10^{-16} centimeter, it is possible that they have a structure smaller than this. But there's another problem: the quarks are confined. We can never isolate them, so, again, what's the use of assuming they are made up of more elementary particles that we also won't be able to see. True, from most indications we will never see quarks and certainly won't be able to see the particles they are made of. But we can detect quarks indirectly, and we may, perhaps, be able to do the same thing for the particles they are made of when larger accelerators become available.

We refer to the particles that make up quarks and leptons by the generic name, prequarks. We will see, though, that within the various theories they are given specific names. There have been many of these theories published over the years but I will only talk about two of them. The first was published in 1974 by Pati and Salam. They referred to their prequarks as preons. In setting up their theory they selected three basic physical properties: electric charge, color, and generation number. Corresponding to each of them they assumed there was a preon family. They called the family corresponding to color, the chromons (see table). There were four chromons—three colored ones and

	Preon	Electric charge	Color	Generation number
Flavons	f_1	+1/2	Colorless	0
	f_2	−1/2	Colorless	0
Chromons	C_R	+1/6	Red	0
	C_Y	+1/6	Yellow	0
	C_B	+1/6	Blue	0
	C_C	−1/2	Colorless	0
Somons	S_1	0	Colorless	1
	S_2	0	Colorless	2
	S_3	0	Colorless	3

a colorless one. The family corresponding to charge was called the flavons. There were two of them, and they had charge $\pm 1/2$. Finally, there were the somons, one corresponding to each of the three generations. To make up a lepton or quark you would select one preon from each family. The electron, for example, consisted of a colorless chromon, a first-generation somon, and a flavon of charge $-1/2$. We see immediately, though, that the charge doesn't come out right; electrons have a charge of -1. Because of this we have to assume the chromons are also charged; if the colorless chromon has a charge of $-1/2$ the electronic charge comes out right. In the same way we can make all other leptons and quarks.

The major difficulty with the theory is that we have selected a family, the flavons, to represent charge, but then we have had to assume the chromons are also charged. Another difficulty is that we always select only one preon from each family. There are no particles that have more than one flavon, or one chromon. Several variations of the theory have been put forward but none of them is entirely successful.

RISHONS

Three years after Pati and Salam published their theory another theory was devised by Haim Harari of the Weizmann Institute of Science in Israel. Harari received his doctorate in physics from the Hebrew University in Jerusalem. Upon graduation he served for four years in the Israeli army, then in 1966 went to the Weizmann Institute and has been there ever since.

Harari decided he wanted a simpler theory than the Pati–Salam one; he disliked several of its features—particularly the problem with charge. Instead of dealing with all three generations at once he decided to concentrate on only the first generation. He called his particles rishons, which means first, or primary, in Hebrew. There were only two of them (compared to Salam's three), but he also introduced antiparticles. He referred

		Electric charge	Color
Rishons	T	+1/3	Red
			Yellow
			Blue
	V	0	Antired
			Antiyellow
			Antiblue
Antirishons	$\bar{T}$	−1/3	Antired
			Antiyellow
			Antiblue
	$\bar{V}$	0	Red
			Yellow
			Blue

to his rishons as T and V, and the corresponding antiparticles as $\bar{T}$ and $\bar{V}$ (see table). T had a charge of $+\frac{1}{3}$, $\bar{T}$ had a charge of $-\frac{1}{3}$ and V and $\bar{V}$ were neutral. The T rishons came in three colors (the same three as quarks) and V in the corresponding anticolors. As we will see later he also had to introduce what he called "hypercolor."

To see how leptons and quarks are made of these rishons, let's begin with the positron. We know it is colorless and has a charge of +1. Trying TTT we find that we get a charge of $\frac{1}{3} + \frac{1}{3} + \frac{1}{3} = 1$, which is exactly what we want. Now for the color. Obviously, if we select three that add up to white we are in business. The recipe for the positron is therefore TTT. In the same way we can show that the recipe for the u quark is TTV. It can be made any color by an appropriate selection of colors. For example, one T could be red, the other blue, and the V could be antired, leaving a blue quark.

There are, however, certain rules that have to be satisfied. First, particles and antiparticles cannot be mixed. There is no physical particle, for example, corresponding to $\bar{T}$VV. Second, there is no particle containing only two rishons; TV, for example, is not a proper recipe.

With the rishon theory, all the first-generation particles and their colors are accounted for. But with only two fundamental particles, all possible combinations were used up and there was nothing left over for the second and third generations. Harari therefore had to consider a different approach. A logical one would be to consider the rishons to be in excited states. This is done in quark theory. But when he worked out the details of such a model he found it didn't work.

There have been several attempts by Harari and others to overcome this problem. The introduction of the Higgs particle, for example, has been suggested as one way around it. The Higgs has no charge or color, and would only add mass, which is exactly what we want, since the muon is nothing but a heavy electron. Another approach is to add rishons in pairs. In such pairs everything except mass cancels. But again this is artificial and seems to destroy the simplicity of the theory. This is, unfortunately, the best we can do at the present time.

If we assume that quarks and leptons are made up of rishons, though, we have another problem: what holds the rishons together? In the case of the quarks we had the color force. And, as might be expected, a similar idea was proposed for this case. Gerard 't Hooft introduced the idea, although he admitted later he was skeptical of the entire theory. He called the new force "hypercolor," and the force carriers, "hypergluons." Like the color force, the hypercolor force would have to be an extremely strong one. Furthermore, it was assumed that it gave rise to confinement. This meant that the rishons were trapped inside the quark, just as quarks are trapped inside protons. The range of confinement in this case is of the order of 10^{-16} centimeter.

It might seem like adding another force of nature is a step in the wrong direction. After all, our object is to reduce the number of forces. But Harari, along with colleague Nathan Seiberg, also of the Weizmann Institute, got around this. They pointed out that the weak force might just be a residue of the hypercolor force—in other words, a van der Waals-type force. As we saw earlier the nuclear force is a residue of the color force. This would mean there are still four basic forces of nature: electromagnetic, gravitation, color, and hypercolor. So far, though, this is little more than speculation.

PROBLEMS

The preon and rishon theories both suffer from the same difficulty: a theory describing their dynamics (how they move) has not yet been formulated. And the reason for this is that there is a serious inconsistency in relation to energy. To see how this comes about let's begin by considering the atom. The total energy of an atom is made up of the mass–energy of its constituents and their kinetic energy. We can easily measure both of these energies. And it turns out that the total energy of the atom is considerably greater than the kinetic energy of its constituents, as expected. This also holds for the nucleus. The nucleus has considerably more energy than the protons and neutrons that are contained within it. When we come to the proton, though, we are starting to see a problem: the energy of the quarks is approximately equal to that of the proton. This is slightly odd, but it is still acceptable. Then we come to the quarks and leptons. If the prequarks are to be confined to a distance of 10^{-16} centimeter (which is required if we aren't able to see them) then they must have a energy greater than 100 GeV. If we take into consideration the possibility that there might also be excited states of prequarks we are talking about energies of several hundred GeV. This compares to only 0 to 5 GeV for the quarks themselves.

How can the constituents of a quark have far more energy than the quark itself? The only way is if the enormous binding energy of the hypercolor forces somehow cancels most of the energy of the constituents. Such cancellations do occur. But when they occur, they are always associated with a symmetry or conservation law. If we are to have cancellation in this case, then we must find the symmetry that causes it. One that has been suggested is called chiral symmetry. Chirality refers to the direction a particle spins—whether it is right-handed or left-handed. And the associated conservation law states that the total number of particles spinning in a given direction going into a reaction must be conserved. In other words, the same number must come out. It turns out that this is satisfied only for massless particles, and therefore if the conservation law is satisfied, the constituent particles must be massless. And this is, of course, what we want. Unfortunately, we have not been able to prove this is the case.

There are many theories in addition to the ones I talked about above. Each tries to show that quarks and leptons are composite in some way. But so far none of them have been successful.

CHAPTER 13

Supergravity

Why, we might ask, are we having so much trouble in our search for a unified theory? Looking at electroweak theory, we see what might be at least a partial answer. Before unification, weak theory was plagued with infinities and was unrenormalizable. But when it was joined with electromagnetic theory the problem mysteriously disappeared. Perhaps we have a similar situation with the grand unified theory. Maybe another field is needed to give a perfect unification. And indeed we have another field, one we have been ignoring up to now—gravity.

Gravity wasn't always ignored, though. Einstein spent the last thirty or so years of his life trying to unify gravity and electromagnetism (the only two fields known when he began working on the problem). But he failed. He failed, according to Gell-Mann, "because he didn't concern himself with quanta. He didn't accept quantum theory."

Einstein's approach to the problem of unification was different from that of most theorists today. He was trying to bring the electromagnetic field into his theory of the gravitational field (general relativity). But he was unable to.

Why is this so difficult? After all, theorists have been confronted by seemingly insurmountable problems before, and solved them. What's so different about this case? There is indeed a difference, and it goes to the foundation of the two theories. Grand unified quantum theories, as the name implies, are based on quantum theory. Forces are a result of the exchange of particles. But in general relativity the force of gravity is not related to a particle exchange. It is the result of the curvature of space–time, a curvature that is caused by the matter within it.

But if general relativity is to be unified with the other forces of nature it obviously has to be quantized. If we did quantize it we would have a version in which gravity, like the other forces of nature, is transferred via virtual particles. We refer to these particles as "gravitons." Over the years many attempts at quantization have been made, but so far none has succeeded. It is, in fact, one of the major problems that now confronts theorists.

Even though we have no such theory, theorists have devised an interaction theory in which gravitons are emitted and absorbed by other particles. Using Feynman diagrams they are able to do simple calculations that give reasonable answers. But in the case of higher order calculations they run into the same problem that plagued the other theories: the appearance of infinities. But in the case of the other theories we were able to renormalize these infinities away. Why not do this here? And indeed, theorists have tried, but the problems are much more serious in this case.

A few years ago, however, a different approach was tried, an approach we now refer to as "supergravity." And it has caused a lot of excitement. "We now have theories that could fulfil Einstein's dream of a unified theory," said Gell-Mann in an interview shortly after supergravity was invented. Freeman Dyson agreed with him. "Supergravity is in my opinion the only extension of Einstein's theory that enhances, rather than diminishes, the beauty and symmetry of the theory. It is the first theory that deserves on aesthetic grounds to be true." There are others, though, who are not quite as sure. "I think supergravity is not the solution," says Ne'eman, "but an interesting side phenomenon." And Yang cautions us, ". . . it is important to remember that physics is not mathematics, just as mathematics is not physics. Nature chooses only a subset of the very beautiful and complex and intricate mathematics that mathematicians develop."

Simply stated, supergravity is an extension of ordinary gravity; instead of one gauge particle, there are many. It is the outgrowth of a type of symmetry referred to as "supersymmetry." One reason theorists are so excited about it is that it includes general relativity theory within it. Moreover, it is a quantum theory. The origins of the theory date to the early 1970s when two groups of Russians, Y. A. Golfand and E. P. Likhtman at the Lebedov Physical Institute in Moscow and D. V. Volkov and V. P. Akalov at the Physical–Technical Institute in Kharkov, formulated the basis of the theory. Similar work was done in the United States by Pierre Ramond and John Schwartz of Caltech and Andre Neveu of the Ecole Normale Superieure in France. It wasn't until 1973, though, that a simple and complete theory was formulated by Julius Wess of Karlsruhe University in Germany and Bruno Zumino of CERN.

Wess and Zumino met at the University of Vienna while Zumino was visiting as a guest lecturer. Wess attended his lectures and was attracted to the work he was doing. "I talked to him and we began working together," said Wess. But then Zumino returned to New York University, where he was teaching at the time. A few weeks later Wess got a letter from Zumino asking him if he would be interested in coming to NYU and continuing the collaboration. Wess was delighted with the invitation, and was soon on his way. "We worked together for several years at NYU," said Wess.

But it wasn't at NYU that the breakthrough in supersymmetry occurred. They were both at CERN. Wess said that he and Zumino had just attended a talk by Shoichi Sakata on the string model in which Sakata talked about certain types of symmetries. "We began to wonder if, perhaps, we could apply such symmetries to particles," said Wess. And so the two of them began to work on a theory.

"It's interesting," said Wess, "Pauli actually thought of the same idea many years before us but he never followed up on it." The idea Wess is referring to is bringing the two types of particles in the universe, namely fermions and bosons, into the same family. Before we talk about how they did this, let's briefly review some of the concepts that we will need in the explanation. First, spin. As we saw earlier, particles spin like tops, or at least we can think of them as spinning like tops (physically, this may not be true but we won't worry about it; the concept works and that's the important thing). If the particle has no spin we say it has spin 0; if it has a small amount we say it has spin $1/2$. If it spins twice as fast we say it has spin 1. Incidentally, it can only spin with these specific values; no values in between are possible.

Physicists have divided particles into two classes according to their spin. If they have spin 0, 1, or 2 they are called bosons; if they have spin $1/2$ or $3/2$ they are called fermions. Bosons are exchange or "force" particles; fermions are what we might call "matter" particles. But fermions and bosons also differ in another respect. Bosons are gregarious; they like to get together, and, in fact, two can occupy the same point in space. Fermions, on the other hand, prefer to keep their distance from one another; they obey the Pauli Exclusion Principle. It is, in fact, because of this principle that atoms exist. Each electron within the atom exists in a different "state."

Now, back to what Wess and Zumino achieved. They discovered that by applying a special type of symmetry (they called it "supersymmetry") they could mathematically change fermions into bosons and vice versa. It is easiest, perhaps, to think of this in analogy to isospin. There we saw that in an imaginary isospin space a neutron could change into a proton. Which of the two particles it was depended on which direction an imaginary arrow was pointed—up or down. Wess and Zumino contructed a similar imaginary "superspace" in which they visualized a "superparticle." If the imaginary arrow was up, the superparticle was a fermion; if it was down, it was a boson.

They soon saw, however, what appeared to be a serious

problem: if they had two bosons occupying the same position in space–time, and transformed them to fermions, the two fermions could not occupy the same position. Examining the mathematics, though, they found that something odd happened: the fermions shifted slightly. Indeed, if they transformed one of the fermions back to a boson it also shifted. In short, repeated application of the fermion–boson transformation moved the particles in space–time.

Wess and Zumino were delighted at the prospect of changing fermions into bosons as no one had ever done this before. It meant that the universe was simpler than previously supposed: there would only be one fundamental type of particle instead of two. But the theory was still incomplete. It was a global theory; in other words, the transformation changed all points by the same amount. But the really important theories in physics, as we saw earlier, are local or gauge theories. Wess and Zumino began to wonder if it would be possible to make the theory into a local one. They soon found that it wasn't an easy problem.

While Zumino continued the search for a local theory, Wess moved on to other problems. Stanley Deser of Brandeis University then joined Zumino in the quest. But others were now starting to get interested; Dan Freedman, Peter Van Nieuwenhuizen of SUNY at Stony Brook, and Sergio Ferrara of CERN also began a search. And, although both groups achieved the goal about the same time (1976), it was the latter group that published first. Priority is still a hotly debated issue and both groups are convinced they were first. I asked Julius Wess who he believed was actually first (and of course expected a slightly biased answer). "I think Zumino and Deser got there first," he said. "But they were excited about their results and didn't publish them while they looked further into the theory." On the other hand, the Zumino–Deser paper referred explicitly to the earlier Freedman–Van Nieuwenhuizen–Ferrara paper. Regardless of who gets there first, though, it is the one publishing first who gets the honor. But in this case there was less than a month difference in publication times.

The first thing to pop out of the local supersymmetry theory, or supergravity theory as it is now called, was the graviton, the exchange particle of the gravitational field. That it appeared naturally within the theory was a significant and important result. Like the photon it was massless, but its spin was unlike that of any other known particle. It was 2. Furthermore, there was something else strange about the theory: a particle that differed from the graviton by spin $1/2$, what we now call a gravitino (it is assumed to have a spin of $3/2$), existed. Other than spin little is known about this gravitino. In the simplest supergravity theory it has zero mass, but in other more complex theories it is massive.

The simplest supergravity theory consists of only gravitons and gravitinos. And of course we know this doesn't correspond to our world. Fortunately, there are other theories, the most important of which are called extended supergravity theories. They are much more restrictive than the nonextended variety and come in only eight versions. They are labeled by the letter N, which corresponds to the number of gravitinos in the theory. In the N = 1 theory there are, again, only the graviton and gravitino. The N = 2 theory has a graviton, two gravitinos, and one particle of spin 1.

Each of the fermions in the N = 1 theory is related through a supersymmetric transformation to a boson, just as the graviton is related to the gravitino. And this seems to imply that the bosons of our world are related via a supersymmetric transformation to the fermions. But this isn't the case. There is, indeed, a particle corresponding to each of the particles in our universe—a "superparticle" that differs by spin $1/2$. These superpartners, as they are usually called, however, are assumed to be presently beyond the energy of our accelerators. The most interesting of the extended theories is the largest one, N = 8. It contains 1 graviton, 8 gravitinos, 28 spin 1 particles, 56 spin $1/2$ particles and 70 spin 0 particles. It is natural, of course, to try to relate these particles to the known particles of the world, but so far we haven't had much luck with this.

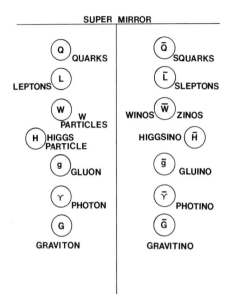

The supermirror—worlds of particles and superparticles.

According to the theory there is a superpartner called the selectron corresponding to the electron. Similarly, there is a squark corresponding to the quark. In these cases the prefix "s" is added; in other cases the suffix "ino" is added. Thus, corresponding to the gluon there is a gluino, and to the photon, there is a photino.

If supersymmetry were a perfect symmetry the mass of the partners would have to be the same as the masses of the known particles. But this obviously can't be the case. Otherwise we would have atoms made of selectrons along with ones made of electrons, and they would have quite different properties. We know of no such atoms. The symmetry must, therefore, be less than exact. In other words, it must be a broken symmetry. This is, of course, the way the W particles get their mass. If it is a broken symmetry then the superpartner can have a larger mass—a mass that is beyond the capability of our present accelerators. This seems to be the most reasonable answer to why we don't see these particles and the one most supertheorists accept.

Because of their huge mass these superparticles do not play a very important role in our present universe, but they may have been important in the very early universe—in the first fraction of a second after the big bang explosion. If we go back in time to this event (back to higher and higher energies) we eventually get to the time when the effects of quantum gravity were important. This is called the Planck era. According to recent theories the universe was quite different during this era. Space and time may have been disconnected—like a churning, foaming froth. At this time, according to the theorists, a single unified force prevailed—a unification of all four known forces of nature. The particles carrying the superforce, the superparticles and the ordinary particles were indistinguishable. The universe was completely symmetric. Then, as the universe cooled, symmetry breaking took place and the superforce broke up. One by one the known forces of nature broke free. And as cooling continued the superparticles gradually disappeared from the universe, leaving only ordinary particles.

How massive are the superparticles? Will we ever see them with accelerators? We still do not know, and we have no reliable way of predicting their mass. Our hopes at the present time are pinned on the Superconducting Supercollider and other large accelerators that are presently being built or will be built in the future. I asked Julius Wess if he thought they would ever be found. "I'm optimistic," he said, nodding his head. "I think they will be. They should be in the trillion electron volt [TeV] region." "But what if they aren't found?" I asked. Would it be the end of supergravity? "Not necessarily," he replied. "We would just have to make some revisions."

Whether or not supergravity turns out to be the correct theory, it has solved an important problem. Earlier I mentioned that when theorists attempted to do calculations using quantum gravity, infinities popped up. But when the same calculations were done using supergravity these infinities disappeared. For every positive infinity associated with a particle, there is a negative infinity associated with its superpartner, and the positive and negative infinities cancel. So far, all calculations have given finite results.

But if supergravity is to be an acceptable theory it is important that we verify it experimentally. The first thing we would like to show, naturally, is that there is a correspondence between the predicted particles and those of our world. And so far we haven't been able to do this. It is possible, of course, that the superparticles will appear when more energetic accelerators are built. Aside from this, though, there are two predictions that have an important bearing on experiment. First, supersymmetric particles must always be produced in pairs. And second, when a supersymmetric particle decays an odd number of supersymmetric particles must appear in the products. This means that after a collision there will eventually only be one supersymmetric particle left—the lightest one. It will remain because there is nothing for it to decay to. The question then is: what is this lightest particle? We do not know for certain but it has generally been assumed to be the photino. This means, then, that we would attempt to detect the photino.

When the details of such interactions were examined, though, it was seen that it would be almost impossible to detect the photino directly. The reason is that it reacts extremely weakly with ordinary matter and would easily escape detection. Energy would therefore appear to be lost. This is, of course, reminiscent of the neutrino. Its existence was predicted because there was a certain amount of missing energy in beta decay. And, of course, we eventually found it. Could we do the same thing for the photino? We could, and indeed several experiments have already been performed.

The two best places to look for the photino are in electron–positron collisions and proton–antiproton collisions. Let's consider electron–positron collisions first. In this case we might expect the creation of a selectron pair, which in turn would decay to an electron pair and two photinos. The two photinos would escape detection and therefore their energy would appear to be missing. Such events have been searched for at PEP (Stanford) and at PETRA in Germany. But so far none have been found.

In the case of proton–antiproton collisions things become much more complicated. The proton is made up of three quarks and a number of gluons, and because of this, many different events are possible. If a quark from the proton collided with a gluon in the antiproton it can be shown that we could expect to see jets of hadrons emanating from the area of the interaction, along with two photinos. Since we would not be able to see the photinos we should look for jets—in this case three are predicted—along with a certain amount of missing mass. There was considerable excitement when events of this type were found at CERN in 1983. But when the data were checked the probability that it was indeed a superparticle appeared to be low.

Many problems remain in supergravity but it still has much going for it; in particular, it includes gravitation (general relativity) and could therefore eventually be the basis of a unified field theory. Certainly it has increased our understanding of how such a theory should be constructed and what properties it should have. I asked Wess if he thought such a unified field theory could ever be achieved. He laughed, then thought for a moment. "It might be an asymptotic process," he said. "But without such motivation we wouldn't try as hard. Supersymmetry and supergravity are needed to make a unified field theory thinkable."

CHAPTER 14

Adding More Dimensions

The shortcomings of supergravity were soon evident, but the theory's mathematical structure was so elegant that few were ready to toss it aside. Many were convinced that a way around the difficulties would eventually be found—a variation in which the problems would miraculously disappear. So far no such variation has appeared, but an important breakthrough has occurred. And most theorists believe that it is an important step in the right direction. The breakthrough I'm referring to centered on an extension of Einstein's general theory of relativity that was published by the mathematician Theodor Kaluza in 1921. The theory was extended by the Swedish physicist Oscar Klein in 1926. It is now called the Kaluza–Klein theory.

KALUZA–KLEIN THEORY

Kaluza, a low-ranking scholar, a "privatdocent," at the University of Königsberg in Russia, became interested in general relativity. He noticed that Einstein's theory, which described the gravitational field, and Maxwell's theory, which described the electromagnetic field, could be brought together within the same framework; in short, they could be unified.

At first sight it might seem strange that two fields with such different properties could be unified. To begin with, the electromagnetic field is 10^{38} times stronger than the gravitational field; furthermore, it is experienced only by charged particles. Gravity, on the other hand, is experienced by all massive particles. The only property that the two have in common is that they both act over long distances.

Kaluza brought the two fields together by adding a dimension to Einstein's theory. Einstein's gravitational field equations are usually written in four dimensions—three of space and one of time. Kaluza wrote them down in five dimensions and found that Maxwell's theory "fell out" of the fifth dimension. He was sure he had unified the two fields and quickly wrote up a paper and sent it to Einstein.

You may wonder why he didn't just send it to a scientific journal. At that time papers were not accepted from unknown scholars. They had to be accompanied by a recommendation from a well-known scientist. And since Kaluza's paper was on general relativity, it is perhaps no surprise that he sent it to Einstein. Today we have "referees" who perform the same task.

Einstein was enthusiastic about Kaluza's theory. Even before he read the paper in detail he wrote back to him. "The idea that the electric field quantities are mutilated . . . has also frequently and persistently haunted me. The idea, however, that this can be achieved through a five-dimensional cylinder-world has never occurred to me and would seem to be altogether new. I like your idea at first sight very much. . . ." Einstein continued by saying that if, upon detailed reading, he found no serious objection he would present the paper to the Prussian Academy as soon as possible.

A week later, after he had studied the paper, he sent a second letter to Kaluza. He was still enthusiastic about it. "I have read your paper and found it really interesting," he wrote. "Nowhere so far can I find an impossibility." But then comes a reservation. "On the other hand, I have to admit that the arguments brought forward so far do not appear convincing enough. I would like to suggest considering the following . . . ." He then suggested that Kaluza show ". . . that geodesic lines [minimum length lines] . . . give the trajectories of electrically charged particles under the simultaneous action of the gravitational and electric field." He ended the letter by saying that if this were done, he would be convinced the paper was correct. There is no indication, however, that Kaluza replied to Einstein's criticism. And this is perhaps why there was a long delay in the publication. Einstein did not present it to the academy, or have it published, for almost two and a half years. Finally, though, in October 1921 he wrote Kaluza. "I am having second thoughts about having restrained you from publishing your idea on a unification of gravitation and electricity two years ago. . . . If you wish I shall present your paper to the academy . . . ."

Later in the year Kaluza's paper appeared in the journal *Sitzungsberichte der Berliner Akademie* with the title, "On the Problem of Unification in Physics." It was the only paper that he ever wrote on the subject. And for a while there was considerable interest in it.

Before 1919 Einstein had spent little time thinking about unification. But Kaluza's paper sparked his interest, and this interest remained throughout the rest of his life. The following year he, along with colleague Jakob Grommer, published a brief paper commenting on Kaluza's theory. It was not an important paper, though, and did not extend the theory. In 1927 he wrote two more short papers on it. Even after he came to the United States several years later he continued struggling with the theory, but never did make a significant contribution to it.

In 1926, however, an important extension was made by Oscar Klein. Einstein was impressed by Klein's paper; writing to Paul Ehrenfest he said, "Klein's paper is beautiful . . . ." Later he wrote to H. Lorentz, "It appears that the union of gravitation and Maxwell's theory is achieved in a completely satisfactory way by the five-dimensional theory."

Klein made several important contributions to the theory. Kaluza had only considered weak fields. Klein showed that the theory applied even if the fields were strong. Furthermore, he brought quantum mechanics into the theory. And he gave a

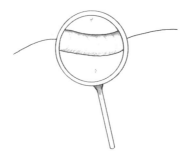

Close-up of a section of a line showing a tiny circle associated with each point of the line. The continuum of circles is a cylinder.

convincing explanation of why the fifth dimension is not observed in nature.

Kaluza had been bothered by the fifth dimension, and felt that its apparent absence in nature should be explained. He explained it in the following way. A line, as I'm sure you know, is nothing more than a continuum of points. Assume that each of these points has a small circle associated with it; the sequence of these circles will give a cylinder, but from a distance it will still look like a line. Kaluza assumed that space had such a cylinder associated with it—this was his fifth dimension. He assumed the cylinder was extremely small and therefore undetectable, but he had no way of calculating its diameter. When Klein introduced quantum theory into the theory he found that he was able to calculate it. And the result he obtained explained why we hadn't seen it. The cylinder had a radius of only 10^{-32} centimeter. This is 10^{16} times smaller than anything we can see at the present time.

Another important property of the Kaluza–Klein theory is its prediction of particles. As we saw above, the fifth dimension is in the form of a cylinder. And since particles in quantum theory are waves, we can associate waves around the circumference of the cylinder with particles. The energy of the particle would depend on its wavelength: the shorter the wavelength, the greater the energy. And since energy is associated with mass, the shorter the wavelength the more massive the particle. The simplest wave—the one with its wavelength equal to the circumference—would therefore correspond to the particle of lowest mass. If two wavelengths fit around the cylinder we would have a particle that was twice as massive. Similarly, if three wavelengths fit, we would have one that was three times as massive. A whole spectrum of particles can be obtained in this way.

The difficulty with this idea comes when we calculate the masses of the particles. They are high. Even the lightest one is 10^{16} times as heavy as the proton, so there is no possibility that we would ever be able to observe them. They may, however, have been present in the early universe.

After Klein's work of 1926 and the papers by Einstein, Kaluza's theory went into decline. No one worked on it, and it was soon considered to be an interesting, but not completely successful unification. Theorists switched their attentions to gauge theories and there was little need for extra dimensions. But eventually the new theories ran into problems and they began to look for a different approach. The Kaluza–Klein theory was pulled from the bookshelves, dusted off, and reexamined. And it was soon shown to have considerable promise.

MODERN THEORIES

The modernization of the Kaluza–Klein theory began in the late 1970s. Kaluza had only included two fields in his unification, the only two that were known at the time: gravity and electromagnetism. But we now know there are two more: the strong and weak nuclear fields. For a complete unified field theory they would have to be included. With this in mind a number of theorists, including Eugene Cremmer, Bernard Julia, and Joel Scherk of the University of Paris, Bryce DeWitt of the University of Texas, and John Schwarz of the California Institute of Technology, tackled the problem.

Because there were two more fields, more dimensions were needed. We are still not certain how many more are required, but as we will see, an eleven-dimensional theory has recently attracted much attention. It was also soon discovered that bosons appeared naturally within the theory, but fermions did not. The only way we could get them was by adding a fermion field to the theory. And this was not satisfactory. There was, fortunately, a way out of this dilemma: if we combined supergravity with the Kaluza–Klein theory, in other words if we add more dimensions to supergravity, both types of particles appeared without having to make any additions.

HIGHER DIMENSIONS

Since the Kaluza–Klein theory is a theory of higher dimensions, let's take a moment to consider these dimensions. The first question that comes to mind is: How do we visualize, or think of, them? Answer: for the most part we can't. We live in a three-dimensional world and it is only possible to visualize up to three dimensions. This doesn't mean that a higher-dimensional space can't exist. It can; in fact, we can easily deal with such a space using mathematics. To see how we might try to visualize them we'll begin with a one-dimensional line. We can use it to generate two dimensions by merely projecting it out in a direction perpendicular to itself. This gives a two-dimensional sheet. To go to three dimensions we project again in a direction perpendicular to the face of the sheet. Thinking of this in reverse we see that two dimensions is the "projection" of three dimensions. Our shadows, for example, are two-dimensional projections of us. A movie projected on a screen is also a two-dimensional projection of a three-dimensional world.

If we wanted to create four space dimensions we would

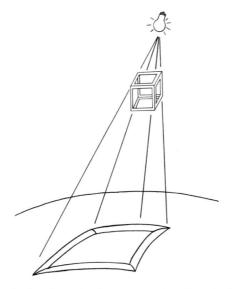

Projection of a three-dimensional square onto a two-dimensional surface.

obviously apply a projection in a direction perpendicular to three-dimensional space. But of course this is impossible. You can easily verify this for yourself by taking four matchsticks and trying to place them so that there is 90 degrees between each of them. As you will see, this is easy to do with three sticks, but impossible with four.

But, as we have seen, there is another possibility: the higher dimension might be invisible. It could be in the form of a tiny cylinder. An object traveling in this dimension would go around and around the cylinder, and nowhere else. We can, in fact, extrapolate this idea to a two-dimensional sheet; each point of the sheet could, for example, have a small circle, or even a small two-dimensional sphere, associated with it. But modern Kaluza–Klein theories deal with up to eleven dimensions. Even if we had a small sphere associated with every point of three-space we would still only have five space dimensions (and one

Three matches showing the three dimensions of space.

of time). What do we do in this case? We'll leave that to the next section.

SUPERGRAVITY

As I mentioned earlier, many of the problems of pure Kaluza–Klein theories are overcome when they are combined with supergravity. The resulting theories can be set up in any number of dimensions but because of a discovery made in 1978 by E. Cremmer and B. Julia it seems most appropriate to formulate them in 11 dimensions. They discovered that the $N = 1$ supergravity in 11 dimensions corresponded to the much more complex $N = 8$ supergravity in 4 dimensions. This was an exciting breakthrough, and it seemed to indicate that 11 was the magic number.

But now we have seven dimensions that are not observed. How do we account for them? As before, we can assume that they are "curled up" in small spheres at each point of four-space. Theorists refer to such dimensions as "compactified." But a sphere is only a two-dimensional object. Where do seven dimensions come in? They would obviously have to be thought of as associated with a seven-dimensional "hypersphere" at each point of space.

It turns out, though, that when we compactify these seven dimensions in the Kaluza–Klein supergravity, the other four also get compactified. In the pure Kaluza–Klein theory we can get around this difficulty by introducing a constant called the cosmological constant that Einstein used in his cosmology to stop the universe from expanding. It allows us to keep the curvature of the universe close to zero, as observed. But with supergravity we are not able to do this.

This, unfortunately, is not the only problem. Another is associated with the neutrino. When we calculate the spectrum of particles that come out of the theory we find neutrinos that spin in both directions, in other words, left-handed and right-handed neutrinos. But in our world we observe only left-handed neutrinos. And there are problems with infinities. Many appear in the theory, and we are still not able to get rid of them.

So, while higher dimensional supergravity theories are no doubt a step in the right direction, by themselves they are not the answer. But perhaps they are part of a larger theory. In the next chapter we will see that they are.

CHAPTER 15

Superstrings: Tying It All Together

The search for a supertheory had again hit a snag. The problems were beginning to look insurmountable. Grand unified theories were incomplete, and some of them gave wrong predictions. And the hope of many scientists—supergravity—now seemed to be in trouble. The boat was sinking, and it appeared that the only way around the difficulties was an entirely different approach.

Was it possible that no such theory existed? Theorists found this difficult to believe. There were many indications that the universe was simple, simple enough that it could be described by a single theory—a single set of equations. What was needed was a theory that contained the strong features of grand unified theories and supergravity but got around their problems. Oddly enough, the beginnings of such a theory had been around for several years, but it had not been taken seriously by most theorists. Only a handful had worked on it. We now refer to it as superstring theory.

Within the last year or so superstring theory has hit the scientific world with a bang. "Superstring theory is now the only game in town," said Steven Weinberg after a recent conference. "It's a very beautiful theory, particularly because of its logical rigidity." Gell-Mann agrees, "Superstring theories . . . I think they are the most likely to work," he said recently.

Is this the theory we've been waiting for? Certainly no one knows for sure. But it does show promise—tremendous promise. And it has gotten theorists excited as never before. A few have even gone as far as saying the discovery of superstring theory is one of the great scientific discoveries of the century—on par with the discovery of quantum mechanics and general relativity. And I won't disagree; this could turn out to be the case. But not everyone is convinced. Sheldon Glashow is one who is not; he compares the rigid faith in superstrings to a faith in God. But despite his and other criticisms superstring theory is emerging as the sleeper of the decade.

Where did it come from? How did it arise? Interestingly, it arose out of a theory, called dual resonance theory, that had nothing whatsoever to do with strings. Dual resonance theory was a spinoff from a work we talked about in an earlier chapter—Regge theory. Regge and others showed in the late 1950s that when a certain plot was made, many particles, in particular hadrons, appeared to lie along straight lines. All the particles along these lines, called Regge trajectories, were assumed to be in the same family. For a while there was a lot of interest in Regge theory. But along came Gell-Mann with different ideas based on group theory and Regge theory fell by the wayside.

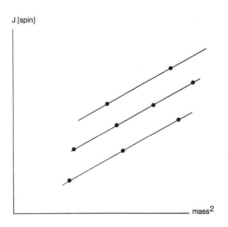

Regge trajectories.

A violin. Note the loops in the strings.

Dual resonance theory was invented by Gabriele Veneziano, now at CERN, in the late 1960s. What it consisted of, essentially, was a formula that enabled theorists to make important predictions about the interactions of particles.

But this formula said nothing about strings. Where, then, did strings come in? The link came in 1970 when Yoichiro Nambu of the University of Chicago, Leonard Susskind of Stanford University, and Holger Nielson of the Niels Bohr Institute in Copenhagen noticed that there was a correspondence between Veneziano's theory and the vibrational states of a string. The best way to think of these vibrational states is to picture them as occurring on a violin string. As you no doubt know, a violin string can vibrate with one, two, or more loops along its length (see figure). Nambu and his colleagues noticed that these vibrational modes could be related to the hadrons. One of the hadrons, for example, corresponded to a string with a single loop, another to one with two loops, and so on. This is not so strange if you stop for a moment and think about it. Particles in quantum mechanics are described by similar vibrational states.

In the case of Nambu's string theory, though, the strings were very special. They had no mass, were elastic, and their ends moved with the velocity of light. But if they had no mass and represented massive particles, where did the particles' mass come from? This was taken care of by the tension of the string: the greater the tension, the greater the mass.

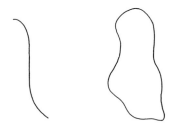

Open and closed strings.

The number of possible vibrations the string could undergo was infinite so the theory could easily describe all the known hadrons. Furthermore, interactions of the particles could be described. Just as particles interact in particle theories so too could the strings that represent particles interact. If two strings came together, for example, they would join to make a single string. In the same way a string could break in two. The details of such interactions were worked out by Stanley Mandelstam of the University of Chicago.

But if the end of one string could join the end of another, the two ends of a single string should also be able to join. And they could. If this occurred we would get a closed string. Thus, we had two types of strings: open and closed. But so far we haven't mentioned particles. Where do quarks, for example, come in? According to Nambu and his group they were attached to the ends of the string. Indeed, if a string broke a quark and an antiquark would appear at the break.

Of course they also had to account for baryons, and baryons were composed of three quarks. They did this by assuming a Y-shaped string—one with three ends. On each end was a quark.

It was an interesting theory. But it had what appeared to be a serious flaw: the only consistent version of the theory required a space of twenty-six dimensions. And most physicists found this unacceptable. After all, our world only has four dimensions in it. Furthermore, there was another difficulty: the theory applied only to bosons. The matter particles of the world—the fermions—weren't included. And to make things even worse, one of the bosons, the lightest one, was a hypothetical particle that moves only at speeds greater than that of light, a particle we call the tachyon.

Progress in overcoming these problems was slow. But it came. The first difficulty was overcome in 1971 when Pierre Ramond now of the University of Florida devised a theory that included fermions. In addition to oscillations, he assumed the string could spin. Shortly thereafter A. Neveu of France and John Schwarz now of Caltech improved the theory and reduced the number of dimensions to ten. Ten was better than twenty-six, but still a serious difficulty. Besides, the tachyons still seemed to be there. And what was perhaps the last straw: some of the particles that were predicted didn't correspond to known particles.

The prospects for the new theory looked dim. But at that time it was just another theory of the strong interactions, and would perhaps be no loss. Its major obstacle was the emergence of a better theory of these interactions: quantum chromodynamics. Since quantum chromodynamics was solving many more of the outstanding problems of the day than string theories were it's obvious which of the two theories was getting the most attention.

But John Schwarz had a strong faith in the string theories. Moreover, he was convinced that they were not just strong interaction theories. He was sure it was possible that they could explain much more. And in 1974, along with Joel Scherk, he published a paper suggesting that string theory could be a Theory of Everything—what we now call a TOE. This meant that it would include gravity. If true, it would indeed be a theory worthy of attention.

Gravity, as we saw earlier, had defied all attempts at unification. The chief difficulty was that it had not been properly quantized. A number of crude and incomplete quantizations

John Schwarz.

had been published and they had been used as the basis of an interaction theory, but when calculations were made using them, infinities appeared. And there seemed to be no way to get rid of them. General relativity had, of course, been incorporated into supergravity, and this had gotten rid of the infinities. But supergravity had other problems—the major one being its prediction of a large number of as yet unobserved particles.

Even with the suggestion that it might be possible to incorporate gravity, the scientific world paid little attention to string theories. Their interests were elsewhere. Supersymmetry and supergravity were coming into their own, and most theorists were jumping on the bandwagon. As far as most of them were concerned supergravity was the best bet for a TOE.

Another important announcement about strings came in 1976. Scherk, along with Ferdinando Gliozzi of the University of Turin and David Olive of Imperial College of London, published a paper showing that it might be possible to incorporate supergravity into string theory, thereby creating a "superstring" theory. And since supergravity contained gravity (i.e., general relativity) superstring theory would also benefit from it. Scherk and his co-workers paved the way, but they didn't follow up on the details. Scherk died tragically shortly thereafter and Olive and Gliozzi went on to other problems. The odds on string theory dropped even lower.

But there was one person still hanging in—John Schwarz. Confident that the theory would eventually blossom, he refused to give up. But although he stubbornly plugged on, he managed to publish nothing. In the summer of 1979 he went to CERN, sure he was the only person in the world working on string theories. But to his surprise he met someone who was as interested in strings as he was—Michael Green. Green had been involved off and on with strings ever since his graduate days at Cambridge. He had, in fact, done his thesis on Veneziano's dual resonance theory. There was no question: they would begin working together immediately. Both were familiar with the problems of string theory and knew that progress would not come easily. Their first problem would be to show that the theory was, indeed, supersymmetric, as Scherk and his colleagues had suggested. Then they had to prove it wasn't full of infinities, as quantum gravity was. But time wasn't on their side. Even though they worked hard throughout the remainder of the summer they came up with next to nothing. Defeated, but far from broken, they left for their respective universities at the end of the summer agreeing to meet again the following year.

And in the summer of 1980 they were again together at Aspen, Colorado. "I never worked so hard, and with such intensity in my life," said Green, referring to work they did that summer. But this time it paid off: they showed that supergravity was indeed contained within string theory. As they had hoped, they had developed a superstring theory. Delighted with their success, they became even more determined to prove the theory was a TOE.

With their success they were sure others would join in. But no one did. This was, in a sense, to their advantage, though; there would be little chance of being scooped. Their next problem was to show that the theory had no infinities. Again they worked with unprecedented intensity—this time for two years. Then, finally, success: there were versions of the theory that contained no infinities. They now had a theory that contained all of the basic fields of nature, including gravity, and predicted all the particles of nature. Furthermore, it had no infinities. You would think that theorists would be eager to jump in. But no. The few who paid any attention to it pointed out that, even though it was infinity-free, it likely had another even more serious problem: anomalies, which can best be described as weird irregularities in the mathematics.

If anomalies did occur conservation laws would not be satisfied, and all sorts of "crazy" things were possible. Negative probabilities might occur. (That would be like saying the probability of you being in an accident tomorrow is minus 20 percent.) Needless to say, if it did have anomalies it would be the end of the theory. But Schwarz and Green weren't finished. They now set out to prove them wrong.

Within a short time they had shown that there were no anomalies in certain restrictive cases. It didn't apply generally, but it was a start. They then examined all the possible groups [SU(4), SU(5), and so on] that could be used as a basis of their theory. One after the other they were tried and rejected. Then they came upon one that worked, called the special orthogonal group [SO(32)]. It was a huge group, but its size was unimportant at this point—what was important was that it worked. Then they found another: a combination of two groups called the exceptional groups ($E_8 \times E_8$). They now had everything they wanted, and at last others began to take an interest.

In August 1984 they made their announcement at Aspen. There was no doubt now: superstring theory was king; it had everything needed to be a TOE. Some of those present thought they were just joking, but when they looked at the successes of

Edward Witten.

the theory they knew it was no joke. Among the most excited at the news was Edward Witten of Princeton University. With only a secondhand account of the proof that it was an anomaly-free theory, he duplicated the feat within an hour. Others at Princeton also soon got into the act. Within the year David Gross, Jeffrey Harvey, Emil Martinec, and Ryan Rohm came up with the "heterotic" string model. It was a closed string model that contained Yang–Mills theory; in other words it was a gauge theory. And as such it is the best theory we have today.

One of the major breakthroughs in string theory came with the realization that the scale was wrong. In early theories the strings were taken to be about 10^{-13} centimeter long. This is the size of the proton, and what we might call a "respectable" length. But when gravity was incorporated into the theory

Schwarz and Green realized that they would have to increase the energy of the fluctuations (which is the same as increasing the tension of the string). The size scale went to the Planck length—10^{-33} centimeter. The strings were then 100 billion billion times smaller than the nucleus. They were to the proton what a speck of dust is to the solar system. They were, in fact, so small that it would be forever impossible for us to see them.

As in the early theory, two types of strings existed in the new theory: open and closed. Also, there were both boson strings and fermion strings. There was a charge at both ends of the open strings, and they could vibrate and spin in an infinite number of ways. The vibrational states included all the massless exchange particles except the graviton. The closed strings could also oscillate, but they had no charge, as they had no end points. The most important closed string, however, the heterotic string, did have charge, but it was spread over the strings themselves.

Waves can run either clockwise or anticlockwise along the closed string. The 10-dimensional theory is associated with the clockwise waves, the 26-dimensional theory with the anticlockwise ones. The fact that we are able to have waves running in either direction is important: it gets around a serious problem. One of the major difficulties of supergravity was its inability to explain the left-hand nature of neutrinos—what we call chirality. With left and right running waves, chirality is built into the theory.

One of the ten dimensions above is, of course, time. But if we include time in our string problem we don't end up with a string; we end up with a string stretched out in time—a sheet. In practice it is the vibrations of this sheet, called a "world sheet," that theorists are interested in. For closed strings we also have a sheet, but it is in the form of a cylinder—an irregular, pulsating cylinder. I should mention at this point that it is only the movement perpendicular to the sheet that we are interested in. One way of picturing these sheets is to think of them as soap films. If there were a wind these films would oscillate in the same way

SUPERSTRINGS: TYING IT ALL TOGETHER

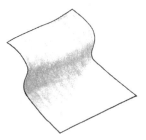

A spacetime sheet. A string stretched out in space–time.

the string sheets oscillate. But there is an important difference between them and our string sheet: the string sheets are in ten dimensions.

But what about particle interactions? After all, this is what particle physics and field theory are all about. Can we describe such interactions in string theory? Indeed we can. In ordinary particle physics we use Feynman diagrams to describe interactions. And we have something similar in superstring theory. The main types of string interactions, you may remember, are the splitting of a string into two strings, and a merging of two strings into one. In the case of closed strings this would amount to the splitting of a cylinder into two smaller cylinders, or the merging of two cylinders. This is, in fact, what our new diagrams look like.

To see the individual strings we merely take a slice through the sheet perpendicular to the time axis (the time axis is to the right in the above diagram). And, as it turns out, the Feynman

A closed string stretched out in space–time.

The merging of two closed strings stretched out in space–time.

diagrams in string theory are generally simpler than the corresponding ones in the point particle theory.

Another problem we've said little about so far is the ten dimensions (or more in some versions of the theory). Our physical world is composed of four dimensions, three of space and one of time. Somehow, we've got to make a correspondence between it and the ten dimensions of string theory. How do we do this? The obvious way is to look back to Kaluza–Klein theory. How did they do it? They "compacted" their extra dimension. We, of course, did the same thing with eleven-dimensional supergravity. In superstring theory six of the ten dimensions would have to be compacted.

Edward Witten of Princeton University was one of the first to work on compactification. He showed that it could lead to what is called a "Calabi–Yau manifold." This is a strange "space" invented by the mathematicians Eugene Calabi of the University of Pennsylvania and Shing-Tung Yau of the University of California at San Diego. Dealing with such strange mathematics was new to most theorists so they recruited Yau to assist them. The question they asked him was: could a correspondence be made between our four-dimensional world and the Calabi–Yau manifold? Yau showed that there were several such manifolds in which this would be possible.

With the problem of how to deal with the extra dimensions out of the way (actually it's not yet completely out of the way) there was still the problem of experimental tests. For a theory to be a "good" theory it has to predict things that are observed, or that might be checked. Can superstring theory predict anything that we can check? So far it hasn't. The major problem is gravity. Because gravity is so weak it is notoriously difficult to devise gravitational interaction experiments that we can perform. And since superstring theory contains gravity we have the same problem. Nevertheless, Philip Candelas of the University of Texas, Witten, and others have come up with a number of possible tests. One of their first predictions was the existence of a strange, nonenergetic, light particle called the axion (it had earlier been predicted in a different way). Experiments that may detect it are now being planned.

Another prediction is "shadow matter." According to one interpretation of the theory there may exist another form of matter in the universe that interacts with ordinary matter only through gravity. There may, in fact, be a "shadow world" that mirrors the ordinary world. We're still not sure how this shadow matter would differ. At the present time, because of the weakness of gravity, there would be no interaction between matter and shadow matter at the particle level. But this may not have always been the case. In the first fraction of a second after the big bang the two types of matter may have reacted violently. According to present ideas both matter and shadow matter were ejected in the explosion, and both should still exist in the universe. There should have been considerable mixing so that galaxies would now consist of both types.

What about our solar system? Would we expect to find any shadow matter here? It is possible, but it would now be hidden. According to calculations, if there were any in either the sun or the Earth it would have settled to the center. We may be living over a core of shadow matter. Scientists are now attempting to devise experiments to see if this is the case.

Not only may there have been shadow matter present in the very early universe, but the dimensions of space-time may have been quite different. If superstring theory is fact, there would have been ten dimensions, all essentially the same. But as expansion occurred, for some reason we do not yet understand, six of the dimensions remained small while the other four expanded with the universe. The six are, in fact, now compact because they were left behind.

To many, superstring theory is a mathematical miracle. They admit that it is far from complete but that it has tremendous potential. Much work is left to be done. It may take decades to understand the theory completely. But this isn't really so bad if we look back at quantum theory. When Schrödinger put forward his psi (ψ) function we had no idea what it was. We may be at a similar stage with superstring theory.

Many problems remain. Why, for example, were six dimensions compactified and not more or less? What is the relationship between strings and the overall universe? What is the role of spontaneous symmetry breaking? How can we test the theory?

While many are confident that we will eventually be able to answer these questions, there are skeptics. Sheldon Glashow, for one, isn't convinced that superstrings are the last word. But as the professor in a recent lecture I attended on superstrings (a friend of Edward Witten) said at the end of the lecture, "the last word on superstrings has not yet been Witten."

CHAPTER 16

Cosmic Strings and Inflation

A book on the unification of particles and fields wouldn't be complete without a discussion of the implications of this unification on the origin of the universe. How do grand unified theories, supergravity, and superstring theory affect the way we believe the universe came into being? As we will see, they do have important implications.

The theory that has for many years been the accepted theory of the origin of the universe is called the big bang theory. It assumes that the universe began in a gigantic explosion about 18 billion years ago, and has been expanding ever since. It is an excellent theory in that it explains much of what we observe. But it is not problem-free. There are, in fact, several difficulties with it.

The first of these difficulties came to light shortly after astronomers began studying the distribution of galaxies in space. They found that, regardless of which way they looked into space, the galaxies were approximately uniformly distributed. Even more amazing was the radiation that was generated in the big bang, radiation we now refer to as cosmic background radiation. It initially had an exceedingly high temperature but as the universe expanded it cooled until it now has a temperature of about 2.7 K. And regardless of which way you looked it had the same temperature. But why is it so uniform? The only way this would be possible is if the explosion that created it was extremely uniform, yet we know from experience that ordinary explosions here on Earth aren't uniform. This seems to imply that there was something "miraculous" about the big bang explosion. And this isn't the only thing that is miraculous about it. Looking closely at the expansion rate of the universe we find a second miracle. To see how this occurs we ask: what will eventually happen to the universe as a result of its expansion? Will it expand forever, or will it eventually stop and collapse back on itself? The answer depends on how much matter there is in it. If there is over a certain critical amount the mutual gravitational pull of the matter will stop the outward expansion and the universe will collapse. If not, it will expand forever. (If it expands forever it is said to be negatively curved; if it collapses back on itself it is positively curved.) It is important, therefore, that we get an accurate estimate of the average density. The first estimates indicated that there was not enough matter in it to stop the expansion. But in these initial attempts many things were overlooked: black holes, small dwarf stars, and so on. Furthermore, we eventually realized that one of the more common particles of the universe, the neutrino, might have mass. As more and more contributions were added to the overall mass of the universe its average density began to approach the amount needed to stop the expansion. We still do not have an exact value for this number, but something amazing appears to be happening: it is getting very close to the critical density. And if it does indeed have this value it will be flat.

This might not seem amazing at first but if you think about it for a moment you realize it is indeed a miracle. After all, it could only occur if the force of the explosion exactly balanced the inward gravitational pull of the matter expelled in the explosion. The explosion could have had any value. But it doesn't. It is finely tuned to the amount of matter in the universe. And as Peebles and Dicke pointed out in the late 1970s this "flatness" is not explained by the big bang.

Another problem with the standard big bang model was pointed out by Wolfgang Rindler in 1956. To understand it let's begin by assuming the universe is 18 billion years old. We still don't know its exact age, but most cosmologists agree that this is in the right ballpark. Now assume that we observe a quasar in the distant reaches of the universe; say it is 10 billion light-years away (a light-year is the distance light travels in one year). The light from this quasar will have taken 10 billion years to get to us. Suppose that we now turn in the opposite direction and find another quasar at an equal distance. The two quasars will then be separated by 20 billion light-years. But our universe is only 18 billion light years old; a light signal traveling from one of the quasars would therefore not yet have reached the second. This means that there is no way the two quasars could have communicated with one another since they were born. Yet they look alike. In fact, all quasars generally look alike. Furthermore, we see that the cosmic background radiation from near one of the quasars has exactly the same temperature (2.7 K) as that near the other. The temperature of this radiation, as we saw earlier, is in fact the same everywhere. But if this is the case something must have "told" it to be the same. Yet this is impossible; the two regions are "causally disconnected"; in other words, they were never "in touch" with one another. This is referred to as the horizon problem, and it is another enigma the big bang theory does not explain.

Earlier we saw that the background radiation is exceedingly uniform. The only way it could have become so uniform is if the big bang explosion was equally uniform. But if it was too uniform we have another problem: where did the galaxies come from? According to most theories they formed from inhomogeneities in the material that was generated in the big bang. And to make things worse it has recently been discovered that the galaxies are not uniformly distributed throughout space; clusters of galaxies seem to be arranged in chains, with large voids between them. This is completely at odds with the uniformity of the background radiation. And again the big bang model does not explain why.

We also have a third problem called the monopole problem. Monopoles are heavy particles that were postulated many years ago by Paul Dirac. He was dissatisfied with the way nature seemed to provide electric monopoles, i.e., particles that carry a single electric charge, yet it did not seem to provide magnetic monopoles. A magnet has a north and a south pole, and if you cut it in half it still has a north and a south pole. In fact, you can continue cutting it but you will never be able to isolate the two poles. Dirac assumed, however, that nature did supply such poles and he constructed a mathematical theory that predicted them.

Scientists looked for Dirac monopoles for years, but never found them. Then with the formulation of the first grand unified theory (GUT) they were again predicted to exist. Gerard 't Hooft and A. M. Polyakov, a Soviet physicist, showed in 1972 that one of the mathematical solutions that came out of GUT represented a magnetic monopole. The theory allowed them to calculate some of its properties, in particular its mass and spin. This set off another series of searches, but again none were found.

Now back to the monopole problem. It arises because 't Hooft and Polyakov predicted that they should be common throughout the universe. Yet we haven't found a single one. Something seems to be wrong and again big bang theory is of little help.

The way out of the above dilemmas came with the introduction of GUT into cosmology.

GRAND UNIFIED THEORY TO THE RESCUE

GUTs only have an effect on the very early universe. So let's look back to this time.

The very early universe is divided into short periods of time referred to as "eras." One of the most important of these eras is, in fact, called the GUT era because of the important role that grand unified theory plays in it. But we will begin our discussion by looking at a later era. Then we will travel back in time to the GUT era and beyond.

Each of the eras was introduced with what is called a "freezing." This is similar to the freezing that takes place in water. What we see at a freezing, or "change of phase" as it is also called, is an abrupt change in the nature of the material. In the case of water we see an abrupt change from liquid to solid as water freezes; similarly when water is heated it goes through another phase change as it changes to steam. In the early stages of the universe we had similar changes of phase, and at each of them there were significant changes in the material that made up the universe.

Let's begin, then, by going back to a time one ten thousandth of a second after the explosion. This may seem like an extremely short interval of time, but on the scale we will be talking about, it is relatively long. At this point what is called the quark era had just come to an end and hadrons were for the first time entering the universe. During the quark era there were no hadrons, only their components, quarks. And at this time they were not confined. Confinement came at the freezing point; quarks and antiquarks came together to form mesons and triplets of quarks came together to form baryons.

In a brief instant all the quarks were gone; all that remained was hadrons. There were, of course, also leptons present at this stage. We might ask, though, if every single one of the quarks was confined. Is it possible that there were some left over? A number of scientists believe this may have been the case and they have searched for these so-called "relic quarks." But so far none have been found.

Traveling back in time further, through the quark era to the time of 10^{-10} second we come to the electroweak era. The energy as we enter this era is 100 GeV, and as in the above case there was a freezing associated with it. Before this time the electromagnetic and weak forces were unified as the electroweak field. At the freezing they split and became distinct. During the electroweak era there were only three distinct forces present: the electroweak, the strong nuclear, and the gravitational. In addition to these fields, though, there were W particles, the gauge particles of the weak interactions. But at the electroweak freezing the W particles quickly disappeared, decaying to muons and neutrinos.

Continuing back through the electroweak era we come finally to the GUT era. The freezing associated with it occurred at 10^{-35} second. The energy at this point is an incredible 10^{15} GeV. The GUT era is one of the most important of these early eras, and as the name implies it is explained by grand unified theory. Prior to the freezing the electroweak and strong nuclear field were unified, but as the energy dropped below 10^{15} GeV the strong field froze and broke away. During this era X particles were present in profusion. At the end of it, however, the energy was too low to produce them and they disappeared.

Finally, at 10^{-43} second after the big bang we get to the Planck era, named for the scientist who introduced the quantum. The energy is now an incredible 10^{19} GeV. We know almost nothing about this era, mainly because we do not have a theory that explains it. General relativity, the theory that tells us how space and time behave at high energies, is no longer valid. We need a quantized version of general relativity to explain this era and it is not yet available. Physicists agree, however, that the geometry of space–time is now strange—space twists and contorts and may be full of wormholes that disappear and reappear.

INFLATION

We know that grand unified theory is useful in telling us what went on just after the big bang, but what about the problems we mentioned earlier? It turns out that grand unified theory solves them. More exactly, a theory called inflation theory devised by Allen Guth of Cornell solves them, but inflation theory is based on grand unified theory.

A colleague, Henry Tye, asked Guth in late 1979 if he would work with him on the monopole problem. But Guth knew almost nothing about cosmology and was reluctant to get involved. Eventually, though, he relented. And within a short time the two men realized that the best way to explain the absence of monopoles was to assume the existence of a "supercooling" in the early stages of the universe. Together they published a paper outlining their idea.

Later in the year Guth began looking into the details of this supercooling. What would it do to the universe, he asked himself? The only answer, it seemed, was that a sudden inflation would occur. The entire universe would dramatically expand—at a rate much faster than that of the standard big bang. Then he began thinking about the problems the big bang theory could not solve. He had heard of the flatness problem and the horizon problem. An inflation such as the one he envisioned might solve these problems. Using SU(5) he worked out the mathematical details of the theory and found that an inflation would indeed solve the problems. He knew immediately he was onto something important.

According to the theory, as the universe cooled it got hung up in a supercooled state, what he referred to as a "false vacuum." This state can be best visualized as a trough bordered on both sides by high barriers (see figure). The true vacuum, in other words the vacuum that now exists in the universe, was assumed to be at a lower energy level (position A in the diagram). While the universe was in this false vacuum it was, according to Guth, subjected to a tremendous repulsive force that caused it to expand at a high rate—the inflation we talked about earlier. Every 10^{-34} second the universe doubled in size. Eventually, of course, it had to get out of this state, and according to Guth it did. It got out by "quantum tunneling" through the barriers on either side of it. As this tunneling occurred gigantic bubbles formed. These bubbles were in the false vacuum, but inside of them was the true vacuum. The inflation finally came to an end when the false vacuum was completely replaced with

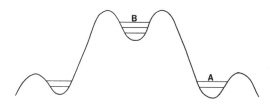

Energy-level diagram showing the false vacuum.

the true vacuum. This occurred about 10^{-33} second after the big bang.

The overall result of inflation was a "dumping" of energy into the universe, a dumping that suddenly increased its temperature to about 10^{27} K.

But as Guth says in his paper, "the inflation scenario seemed to lead to some unacceptable consequences." His major problem was finding a smooth ending to the expansion. Guth realized that the decay of the false vacuum would not occur uniformly throughout space, just as ice does not form uniformly on a pond. If you take the time to watch the ice form, you see that it forms in patches. The patches then grow and join until the entire area is solid. Much the same thing would happen in the case of the universe. The universe would become highly irregular and this would not allow a smooth ending to inflation. Furthermore, in Guth's model it seemed that inflation would not last long enough.

I should also mention that Guth also realized that there was no proof that the universe entered a false vacuum. He only considered what would happen if it did.

For two years Guth's theory stood as an interesting and perhaps brilliant innovation, but unfortunately, a flawed one. The inflation that he described could not have occurred in our universe. But then came the "new" inflation theory. It was put forward by A. Linde of the Soviet Union, and independently by Andy Albrecht and Paul Steinbrandt of the University of Pennsylvania. In the new inflation theory the vacuum is not forced to quantum tunnel to the new lower level and because of this it was slightly delayed and consequently came to a smoother ending. Since this theory was put forward there have been many variations on it. One is called supersymmetric inflation, another is called primordial inflation. There are also models based on supergravity. And recently Joseph Silk of the University of California and Michael Turner of Fermilab have proposed that there was not one inflation in the early universe, but two. They believe that a single inflation does not account for many of the features of the universe on a large scale—in particular, the chains of galaxies.

Now we have to explain how inflation theory explains the problems of the standard big bang theory. First, let's consider the horizon problem. During inflation the universe underwent a rapid inflation that lasted from about 10^{-35} seconds to about 10^{-33} seconds. During this time an increase in size by a factor of about 10^{50} occurred. We know, of course, that the universe was initially causally connected, in other words each point of it was in communication with each other point. It became causally disconnected in the standard model primarily because of the relatively slow expansion. But in the inflation theory we have a rapid expansion—so rapid that it would be virtually impossible for it to become causally disconnected. Thus, inflation solves the horizon problem.

Now for the flatness problem. This problem is, in a sense, linked to the above one. If one is solved so is the other. To see how consider a bug on the surface of a balloon. When the balloon is small it can quickly crawl around it and can easily determine that it is curved. With inflation, though, the radius increases so fast that the bug can no longer detect the curvature. To it the universe is flat. In fact, regardless of the density the universe had before the inflation, it would be essentially flat after inflation.

It's interesting that at the time Guth published his paper there was considerable doubt as to whether a flatness problem actually existed. To emphasize that there was indeed a problem,

Guth added an appendix that began, "This appendix is added in the hope that some skeptics can be convinced that the flatness problem is real."

Third we have the monopole problem. This is the problem that Guth set out initially to solve. To understand his solution we have first to look at how monopoles were created in the early universe. When the bubbles formed, the vacuum field within them was oriented in a certain direction. Furthermore, in Guth's original theory the bubbles were small and many emerged. And when they collided knots formed along the joining surfaces. These knots were monopoles. This means that monopoles were created at each merger and according to calculations mergers were common so monopoles would also have been common. In Guth's model the universe resulted from the merger of numerous bubbles, but in the new theory individual bubbles expanded so fast that our universe is considered to be within a single bubble. It would therefore contain few if any monopoles and we would not expect to see any. In short, new inflation solves the monopole problem because we see only part of one enormous bubble.

There is, of course, another problem—one we have not yet mentioned—the formation of galaxies. It is not solved by inflation theory but many inroads have been made. Much of the work in this area in recent years centers around cosmic strings. Let's look at them.

COSMIC STRINGS

One of the major problems of cosmology is why, if the original big bang explosion was smooth, the universe is now "lumpy." The lumpiness I'm referring to is the inhomogeneous distribution of galaxies in long chains with giant voids between them. It seems strange that the radiation in the universe is so smoothly distributed yet the matter is not. What caused this? The answer may be cosmic strings. Just as monopoles were formed as the universe went through a freezing, so too were cosmic strings. According to recent theories we can think of them as forming along the regions where the bubbles merge. And like monopoles they would also be bizarre objects. They would be so massive that a segment an inch long would weigh millions of tons, yet they would stretch across the entire universe. They would likely exist initially in huge tangled masses, but they are under tremendous tension and kinks would therefore tend to straighten out. But occasionally strings would cross and when they did they would weld together forming huge loops. All such strings, whether straight or in the form of loops, would be vibrating.

Edward Witten of Princeton University had been in on most of the development of superstring theory. And in 1985 he turned his attentions to cosmic strings. He was initially interested in how they might be observed. It had been pointed out in 1984 by Kaiser and Stebbins that cosmic strings would have left their signature on the microwave background, and this signature might be detectable. What Witten discovered while examining the mathematical properties of the strings is that they might be superconductors. In other words, once a current got started along them it would flow forever. If this did occur Witten realized that a tremendous current—up to 10^{18} amps—could be induced in the strings. And since the strings would be vibrating they would emit electromagnetic radiation. This is, in fact, the way electromagnetic signals—for example, those emitted by a radio or TV station—are generated. A current is sent to an antenna where it is caused to oscillate.

In the case of the radiation given off by the string, though, it would be so high—10,000 times as great as that for a quasar—that it would make the entire string glow. We have, in fact, seen glowing threads near the core of our galaxy. Mark Morris of UCLA and Farhid Yusif-Zambal discovered them in 1986.

When Jeremiah Ostriker, also of Princeton, heard of Witten's work he immediately joined forces with him in hopes of producing a model that would predict the lumpiness of the universe. What they produced was a model in which the radiation emitted by the strings pushed matter away from it, creating a void or bubble around it. A large number of such strings would produce chains of galaxies in the regions between the voids where the matter was compressed. The theory may be an important breakthrough. "Superconducting strings may . . . transform our view of the large-scale universe," says Ostriker. These strings may, in fact, explode. When the current reaches a certain limit particles will be emitted along it which will decay in a powerful burst of energy. And we may eventually be able to detect these bursts.

So, while we have not yet observed cosmic strings, they will no doubt play an important role in cosmology for some time to come.

THE ULTIMATE QUESTION

We have been talking about an inflation that self-produced most of the energy of the universe. But we still haven't answered the question: where did the universe itself come from? We do know something of what happened before inflation. I mentioned earlier that the Planck era came before inflation. But as we pass into this era we find we cannot adequately describe it. To do so would require a theory of quantum gravity, and it is not yet available. There is speculation, though, that at this stage space–time was completely different than it now is. It was in the form of a disconnected froth. Furthermore, space may also have had more dimensions than it now has.

But what happened at the beginning of the Planck era to bring the universe into being? Where did the universe come from? Is it possible that it emerged from "nothing"? The difficulty with this is that we find it virtually impossible to visualize what nothing is. A number of scientists, including Villikin, Guth, Coleman, and Pagels, however, have recently considered this possibility. The first to consider the question of how the universe began, however, did not think of it as arising from nothing. Edward Tryon of Hunter College in New York considered the possibility in the early 1970s that the universe arose from a quantum fluctuation of the vacuum. But he did not develop his idea into a mathematically consistent theory and little interest was paid to it at the time.

In time, though, others began to take a serious look at the problem. The most difficult question to answer was: how do we get something from nothing? Coleman, and more recently Pagels, have suggested that "quantum tunneling" may be the way. This tunneling would be similar to the tunneling that took place in Guth's inflationary model. There we had the false vacuum tunneling through to the true vacuum. In this case we would have "nothing" tunneling through to create space–time.

In summary, assuming the above scenario is correct, we have the universe spontaneously erupting out of "nothing." Soon after it appeared it found itself in a false vacuum state. But this state was unstable and a sudden but short-lived inflation occurred. This inflation provided the universe with the energy it needed both to continue its expansion and to create particles. As inflation ceased the universe continued to expand at its usual rate. It then evolved as freezing after freezing occurred until finally the universe we know today was created.

CHAPTER 17

Epilogue

In our journey through the world of elementary particles we have met many different types of particles: muons, pions, quarks, preons, superparticles, and technicolor particles, to name only a few. We have also seen that the breakthrough in unifying the fundamental fields of nature was the realization that they had to be gauge theories. Furthermore, we have looked at the attempts that have been made using higher dimensions, strings, and other strange concepts, to bring these fields and particles together into a supertheory.

Now that we have a better picture of the tremendous strides that physicists have made, it is natural to ask: Where do we go from here? In particular, is superstring theory the supertheory that scientists have been searching for? Certainly, we do not yet know. Only time will give us the answer to that. Superstring theory does have a lot going for it, but as with most other theories there are problems. One of the major ones relates to its predictions; they are, it seems, beyond experiment. This is not the way physics usually works. In the past, experiment and theory have always gone hand-in-hand. Something is predicted and experimentalists look for it; or something is discovered and theorists try to explain it. With superstring theory, and other modern theories, this no longer seems to be the case. We may, in a few years, be able to check on some of the predictions of supergravity, technicolor theory, and GUTs—particularly if the superconducting supercollider (SSC) is built. But there may be things we will never be able to check.

In spite of the tremendous progress that has been made questions remain. And if past experiences are any indication when the answers finally come there will be surprises, surprises that may start us off on an entirely different track. But this is, after all, one of the most exciting things about science. It is unpredictable and at times full of surprises.

One thing about physics, however, is predictable. And this is that the younger generation is always the most enthusiastic about the new theories. They are the ones at the present time who are most convinced that we are in the final stages of achieving a supertheory. Scientists of the older school—Schwinger, Feynman, and Glashow, for example—are not as convinced that this is necessarily the case. We saw this with Einstein. As a youth he made breakthroughs that were difficult for most scientists to accept, but later in life found it difficult to accept new advances such as quantum mechanics.

Despite the skepticism of the older generation, there is an air of optimism that we are close. Only time will tell how close.

Glossary

Antiparticle Corresponding to every type of particle there is an antiparticle. When a particle and an antiparticle meet they annihilate one another, with the release of energy.

Asymptotic freedom The decrease in force between quarks at short distances.

Balmer series A series of spectral lines that arises in hydrogen.

Baryon A heavy particle. Made up of three quarks.

Baryon number (B) A quantum number. All baryons have B = 1. Other particles have B = 0.

Beta decay The decay of the neutron to an electron, a proton and a neutrino. A consequence of the weak interactions.

Boson Particle with integral spin.

Bubble chamber A device used to render the tracks of particles visible. Fluid boils along the path of the particle, creating bubbles that can be photographed.

Charm A property of matter that occurs when it contains charmed quarks.

Charmonium (J/ψ) A system consisting of a charm quark and an anticharm quark.

Charmonium spectrum The energy-level spectrum of charmonium. A representation of the energy levels.

Chirality Having to do with left-handedness or right-handedness of a system.

Chromoelectric line A line representing the chromoelectric force between two quarks. Analogous to electric force in QED.

Chromomagnetic line   A line representing the chromomagnetic force between two quarks. Analogous to magnetic force in QED.

Color   Quality of quarks, like electric charge. The color force is the force of attraction between quarks.

Compactification   A process in which certain dimensions of space–time are made small.

Confinement   The "trapping" of quarks inside hadrons.

Conservation laws   Rules stating that some quantities do not change in an interaction.

Cosmic rays   Charged particles—mostly protons—from outer space.

Coupling constant   A number that represents the strength of a force.

Cross section   The "target area" an incoming particle in an interaction sees.

Decuplet   An array of ten particles. Used in eightfold method.

Dee   One half of a cyclotron. Shaped like the letter D.

Eightfold way   A method of classifying particles into families of eight based on group theory [SU(3)].

Elastic scattering   Particle collisions in which particles do not change their properties.

Electric field   Field that occurs around an electric charge.

Electromagnetism   One of the forces of nature. Exchange particle is the photon.

Electron   The lightest massive elementary particle.

Electron volt   The amount of energy an electron acquires as it moves through a potential difference of 1 volt.

Electroweak theory   The unified gauge theory of the electromagnetic and weak fields.

Exchange particle (gauge particle)   Particles that are passed back and forth creating a force. Exchange particle of the electromagnetic force is the photon.

Family of particles   A group of particles that have approximately the same properties.
Fermion   Particles with a half-integral spin ($1/2$, 1, ...).
Feynman diagram   A diagram that depicts in a simple way how particles interact.
Field   A region of space–time in which a quantity (e.g., electric potential) is specified at each point.
Fission   A splitting of the nucleus.
Flavor   Name given that specifies the type of quark (e.g., up, down).
Frequency   The number of vibrations per second.
Gauge   A measure. A gauge theory is a theory that is invariant under certain symmetry operations.
GeV   A unit of energy. A billion electron volts.
Global gauge invariance   An invariance (maintenance of symmetry) that results when the same change is made everywhere.
Glueball   A neutral meson consisting only of gluons.
Gluon   The exchange particle of the strong interactions.
Grand Unified Theory (GUT)   The theory in which the strong, weak, and electromagnetic forces are unified.
Gravitino   A particle predicted by supergravity. An exchange particle of spin $3/2$.
Group theory   A mathematical technique that concerns itself with collections of elements and how they change under various operations.
h   Planck's constant. Related to quantum theory.
Hadron   Class of particles made up of baryons and mesons. Particles that participate in the strong interactions.
Heterotic string model   A closed string model that is locally gauge invariant.
Higgs boson   The particle that gives mass to the W and Z particles.
Higgs mechanism   The process in which W particles gain mass as a result of the absorption of Higgs bosons.

Inelastic scattering   Particle collision in which particles change their identities.
Inflation   A sudden expansion of the universe.
Infrared slavery   A term describing the confinement of quarks and gluons to the interior of hadrons.
Invariance   Remain the same. A maintenance of symmetry.
Isospin   Corresponds to different orientations in an abstract charge space. A mathematically convenient way of representing the charge states of a family of particles.
Jet   A "spray" of hadrons produced in particle reaction that are all traveling in generally the same direction.
Lamb shift   A slight shift in one of the energy levels of hydrogen caused by self-energy effects.
Lepton   The "light" particle of the universe. Not affected by strong interactions. The electron, muon, tau and their neutrinos.
Local gauge invariance   An invariance that results when random changes are made from point to point.
Magnetic field   The field that is associated with a magnet.
Matrix mechanics   A form of quantum mechanics in which matrices (arrays of numbers) are used.
Meson   A medium-weight particle. Made up of a quark and an antiquark.
MeV   A unit of energy. A million electron volts.
Mixing angle parameter   A parameter used in electroweak theory that relates to how electromagnetism and weak theory are joined.
Multiplet   Group or family of particles. Occurs in group theory.
Muon   A heavy electron.
Neutral current   A weak interaction in which no charge is exchanged.
Neutrino   A particle that is believed to be massless, that is electrically neutral, and experiences only weak interactions.
Neutron   The neutral particle of the nucleus.

GLOSSARY

Parity Pertaining to whether the mirror reflection of a process is the same.

Pauli Exclusion Principle Principle stating that no two fermions with identical properties can occupy the small region of space.

Perturbation series A series of mathematical expressions that arises in interaction theories.

Pion The pi meson. Made up of a quark and an antiquark.

Planck's constant Basic constant of quantum theory.

Positron The antiparticle of the electron.

Preons Hypothetical building blocks, or constituents of quarks.

Quantum A small discrete amount of energy that is absorbed or emitted in particle interactions.

Quantum chromodynamics (QCD) Quantum field theory of the color force. Describes interactions of quarks and gluons.

Quantum electrodynamics (QED) Quantum theory of the electromagnetic interactions.

Quantum numbers Numbers that specify quantized physical quantities such as spin and momentum.

Quark Elementary particle. Comes in six flavors and three colors. Constituent of hadrons.

Quark flavor Refers to type of quark (e.g., up, down).

Renormalization A process in which apparent infinities in various interaction calculations are eliminated.

Renormalization group A mathematical technique developed to relate the structure of QED at one energy to that at another energy.

Representation A way of representing groups of particles. Occurs in group theory.

Resonance A short-lived particle.

Rishon Hypothetical particle that makes up quarks.

Scaling A scale invariance in particle interactions.

Self-energy A back-reaction of a particle on itself.

Spectral lines Bright (or dark) lines that occur when light is passed through an instrument called a spectroscope.

Spectroscopy A study of spectral lines.
Spin Property of an elementary particle similar to spin of a spinning top.
Spontaneous symmetry breaking A process in which a symmetry that holds at a high temperature disappears from the system at lower temperatures.
Standard model Electroweak theory and QCD taken together.
Strangeness A quality that matter possesses if it contains strange quarks.
Strangeness number (S) A quantum number related to the number of strange quarks a particle possesses.
String theory A theory in which elementary particles are assumed to consist of tiny strings.
Strong interactions Interactions associated with the strong or color force. Mediated by gluons.
SU(2), SU(3) Special groups. SU(2) is the special unitary group of 2 by 2 arrays.
SU(5) The special unitary group of five dimensions (or 5 by 5 arrays). The basic group of the Georgi–Glashow GUT.
Superconducting theory The theory of superconductivity—a process in which electricity flows with almost no resistance.
Supergravity A gauge theory of gravitation. An extension of ordinary gravitational theory (general relativity).
Superparticle Particle predicted by supergravity.
Superpartner Partners of superparticles.
Supersymmetry A symmetry in which bosons and fermions are two states of the same particle.
Symmetry Refers to quantities that remain the same after certain operations are performed on them.
Synchrotron An accelerator in which magnetic fields and accelerations are synchronized to keep the particle at a particular radius.
Tau lepton The heaviest known lepton.
Tensor field A field needing ten numbers to specify it.
Uncertainty Principle Principle stating it is impossible to measure the position and momentum of a particle at the same time.

Unified field theory A theory in which different interactions are shown to be identical at a deeper level.
Upsilon A very massive meson composed of a bottom quark and an antibottom quark.
Vacuum polarization A change in the properties of space around an electrically charged particle.
Vector field A field that can be specified by four numbers.
Virtual particle A particle that only lives a short time (restricted by the uncertainty principle). All forces are transmitted via virtual particles.
Weak interactions Feeble, short-ranged interactions. Exchange particles are W and Z.
Weinberg–Salam theory The theory of the unified electromagnetic and weak interactions.
W particle Exchange particle of the weak interactions.
X particle Extremely massive exchange particle predicted by GUT. They allow quarks to change into leptons and vice versa.

Further Reading

Asimov, Isaac, *The Collapsing Universe* (New York: Simon and Schuster, 1977).
Calder, Nigel, *The Key to the Universe* (New York: Viking, 1977).
Close, Frank, *The Cosmic Onion* (London: Heinemann, 1983).
Davies, Paul, *Superforce* (New York: Simon and Schuster, 1984).
Feinberg, Gerald, *What is the World Made of?* (New York: Doubleday, 1977).
Fritzsch, Harald, *Quarks* (New York: Basic Books, 1983).
Nambu, Yoichiro, *Quarks* (Singapore: World Scientific, 1981).
Pagels, Heinz, *The Cosmic Code* (New York: Simon and Schuster, 1983).
Pagels, Heinz, *Perfect Symmetry* (New York: Simon and Schuster, 1985).
Parker, Barry, *Einstein's Dream* (New York: Plenum, 1986).
Pickering, Andrew, *Constructing Quarks* (Chicago: University of Chicago Press, 1984).
Polkinghorne, John, *The Particle Play* (San Francisco: Freeman, 1980).
Silk, Joseph, *The Big Bang* (San Francisco: Freeman, 1980).
Trefil, James, *From Atoms to Quarks* (New York: Scribners, 1977).
Trefil, James, *The Moment of Creation* (New York: Scribners, 1983).
Weinberg, Steven, *The First Three Minutes* (New York: Basic Books, 1977).

Index

Accelerator, 5–14, 49–57
Alpha particle, 2, 21
Alvarez, Luis, 53–54
Ambler, Ernst, 132
Anderson, Carl, 36, 44, 79
Anderson, Phillip, 142
Anomalous magnetic moment, 86
Antineutrino, 41
Antiparticle, 36
Antiquark, 3, 99
Antirishons, 222
Applequist, Thomas, 182
Asymptotic freedom, 156–158
Atom, 1, 18
Axial vector, 134
Axion, 259

B field, 124
Background radiation, 263
Bag, 195
Bag model, 163
Balmer series, 26
Baryon, 63, 151, 250
Becker, Ulrich, 171, 174
Becquerel, Antoine, 125
Beta decay, 39, 129, 135–136
Bethe, Hans, 84
Big bang, 4, 261
Bjorken, James, 105, 112, 115
Black body curve, 22–23

Bloch, Felix, 34
Bludman, S. A., 136
Bohr atom, 25–29
Bohr, Niels, 24–29, 81, 86–87
Bootstrap theory, 106
Born, Max, 32
Boson, 149
Boson strings, 256
Bubble chamber, 57–58

Calabi, Eugene, 258
Calabi–Yau manifold, 258
Candilas, Philip, 259
Causally disconnected, 263
CERN, 6, 9, 13, 166–169, 193–201
Chadwick, J., 37
Chandrasekhar, S., 123
Change of phase, 265
Charm, 104–106, 168–175
Charmonium spectrum, 182–183
Chen, Min, 170–171, 174
Chew, G., 95
Chiral symmetry, 225
Chirality, 256
Chromoelectric lines, 162
Chromomagnetic lines, 162
Chromons, 220
Cline, David, 197
Closed string, 250
Cockcroft, John, 5, 49

INDEX

Coleman, Sidney, 155
Color, 3, 106, 152, 158–160
Color theory, 149–152
Confinement, 153, 160–164
Continuous symmetry, 118
Cosmic rays, 44
Cosmic strings, 270–272
Cowan, Clyde, 41
Cremmer, E., 244
Critical density, 262, 269
Cronin, J. W., 133
Cross section, 61
Cyclotron, 5, 50–57

Dalton, John, 15–17
Davisson, Clinton, 29
de Broglie, Louie, 27–29
Decuplet, 72
Delta particle, 63, 150
Deser, Stanley, 231
Detector, 57
Dirac, Paul, 35, 77–79
Dirac's equation, 78
DORIS, 187–189
Doublet, 65
Dyson, F., 89

Eddington, A., 83
Ehrenfest, Paul, 239
Eightfold method, 69–70
Einstein, Albert, 13, 23, 237–239
Elastic scattering, 109
Electromagnetic force, 4, 43, 237
Electron, 1, 18–19
Electroweak era, 265–266
Electroweak theory, 143–147, 205
Ellis, John, 193–194
Eras, 264
Exceptional group, 254
Exchange force, 42
Exclusion principle, 149
Extended supergravity, 232

Fairbank, William, 101
False vacuum, 267
Faraday, Michael, 16
Fermi, Enrico, 41, 61, 127–130
Fermilab, 8–12, 198–199
Fermion strings, 256
Fermion, 149
Ferrara, Sergio, 231
Feynman diagram, 90–91
Feynman, Richard, 87–92, 113–115, 134
Fine structure constant, 83
Fitch, V. L., 133
Fermi–Yang model, 68
Fission, 128
Flatness, 262
Flavon, 220
Frautschi, S., 95
Freedman, Daniel, 231
Freezing, 265
Freon, 166
Fritzsch, Harald, 152

Gaillard, Mary, 193–194
Gargemelle, 166
Gauge invariance, 121–122
Gauge particle, 124
Gauge symmetry, 120
Geiger, Hans, 21
Gell-Mann, Murray, 2, 66–75, 97–101, 115, 152–153, 247
General relativity, 227–228, 252
Georgi, Howard, 207–214
Glaser, Don, 57–58
Glashow, Sheldon, 103–105, 136, 138, 207–209
Gliozzi, Ferdinando, 252
Gluino, 233
Gluon, 115, 156
Goldhaber, Gerson, 184
Goldhaber, Maurice, 73
Goldstone boson, 142
Goldstone, Jeffrey, 142

INDEX 289

Golfand, Y. A., 229
Gravitino, 232
Graviton, 228
Gravity, 4
Green, Michael, 253–254
Greenberg, Oscar, 150
Grommer, Jakob, 239
Gross, David, 155
Group theory, 68
GUT era, 266
Guth, Allen, 266

Hadrons, 63, 99
Han, Moo-Young, 150
Harari, Haim, 221–223
Harvey, Jeffrey, 255
Heisenberg, Werner, 32–34, 63, 79, 95
Heterotic string model, 255
Higgs boson, 142
Higgs mechanism, 203
Higgs particle, 203
Higgs, Peter, 142
Horizon problem, 263, 269
Houston, W., 81
Hseih, Y. M., 81
Hydrogen atom, 26
Hypercolor, 223
Hypergluon, 223

Iliopoulos, John, 169
Inelastic scattering, 110
Inflation theory, 266–270
Isospin, 64, 124
ISR, 194

J/ψ particle, 2, 180–182
Jet, 191–195
Julia, B., 241, 244

Kaluza, Theodor, 237–241
Kaluza–Klein theory, 237–242
Kaon, 47, 68, 74

Kemmer, N., 138
Kibble, Thomas, 144
Klein, Oskar, 135, 237–241
Kolar experiment, 216
Kramers, H. A., 84
Kroll, N., 84

Lamb, Willis, 6, 81–84
Lamb shift, 83, 86
Langevin, Paul, 29
Lawrence Radiation Lab, 51
Lawrence, Ernest, 5, 50–54
Lederman, Leon, 188–189
Lee, Tsung-Dao, 131–136
Leitner, Jack, 73
Lepton family, 3, 63
Likhtman, E. P., 229
Linde, A., 268
Low, Francis, 153
Lorentz, H., 239

Maiani, Luciano, 169
Mandelstam, Stanley, 250
Marsden, Ernest, 21
Marshak, Robert, 134
Martinec, Emil, 255
Messiah, A., 150
Meson, 42, 63, 100, 151
Millikan, Robert, 100
Millikan oil drop experiment, 100
Mills, Robert, 124
Mixing angle, 138
Mixing parameter, 205
Momentum, 120
Monopole problem, 263
Morris, Mark, 271
Multiplet, 66
Muon, 3, 41, 46
Musset, Paul, 166–169

Naked charge, 157
Naked charm, 157, 183–186

INDEX

Nambu, Yoichiro, 106, 141–142, 150, 162, 249
Ne'eman, Yuval, 71–72, 75
Neutral currents, 137, 167–169
Neutrino, 3, 38, 40–41, 186
Neutron, 1, 37
New inflation theory, 268
Nielson, Holger, 249
Noether, Emmy, 118–119
Nucleon, 64
Nucleus, 21

Okubo, Susumu, 70
Olive, David, 252
Omega-minus, 73–75
O'Neill, Gerard, 175
Oppenheimer, Robert, 78
Ostriker, Jeremiah, 271

Palmer, Robert, 167–169
Parastatistics, 150
Parity, 130–140
Parton, 114–115
Pati, Jogesh, 206
Pati–Salam theory, 206–208
Pauli, Wolfgang, 38–39, 41, 64, 79, 126, 140, 149
Pennsylvania conference, 89
Peirls, R., 140
Perl, Martin, 186
Perturbation theory, 91
Petermann, A., 153
PETRA, 193
Pierre, Francois, 184
Photino, 233
Photoelectric effect, 23
Pion, 46, 64, 143
Planck era, 234, 266
Planck, Max, 21–22
Plum pudding model, 18, 21
Politzer, David, 154–156, 182
Positron, 3, 36
Powell, Cecil, 46

Preon, 219–224
Probability wave, 32
Proton decay, 214–216
Proton, 1
Psi, 31

Quanta, 23
Quantum chromodynamics (QCD), 152, 158–160, 251
Quantum electrodynamics (QED), 79, 122, 124–125
Quantum field theory, 65
Quantum mechanics, 29–37
Quantum number, 64
Quantum tunneling, 267
Quark, 2, 97–116
Quinn, Helen, 210–212

Rabi, I. I., 81
Ratio R, 185
Regge, T., 95
Regge theory, 95
Regge trajectory, 96
Reines, Fredrick, 41, 209–210
Relic quarks, 265
Renormalization, 81, 84, 92
Renormalization group, 153–154
Representation, 69
Resonance, 62, 103
Richter, Burton, 175–181
Rindler, Wolfgang, 262
Rishon, 221–224
Roentgen, Wilhelm, 125
Rohm, Ryan, 255
Ross, Graham, 193–194
Rubbia, Carlo, 168, 197–201
Rutherford, Ernest, 2, 19–21

S matrix, 95, 106
Sakata, Shiochi, 68, 97
Sakharov, Andrei, 213
Salam, Abdus, 71, 137–140, 206–207
Samios, N., 73

Scalar field, 133
Scaling, 113, 153
Scattering experiment, 61
Scherk, Joel, 242, 251
Schrödinger, Erwin, 30–31, 77
Schwartz, John, 242, 251–254
Schweber, Silvan, 105
Schwinger, Julian, 85–86, 135, 142
Seiberg, Nathan, 223
Selectron, 233
Self-energy, 79
Serber, Robert, 98
Shadow matter, 259
Shadow world, 259
Shelter Island conference, 83–85, 92
Sigma particle, 66
Silk, Joseph, 269
SLAC, 57, 107, 111
Slavery, 158
Somon, 220
SPEAR, 186–189
SPEAR experiment, 175–181
Special orthogonal group, 254
Spectral lines, 18–19
Spin, 63, 230
Standard model, 205
Stanford experiment, 2
Stochastic cooling, 198
Strange particles, 59–61, 67–68
Strings, 14
String theory, 248–251
Strong nuclear force, 4, 124
Steuckelberg, Ernst, 94, 153
SU(5), 212–214
Sudarshan, E. C., 134
Superconducting supercollider (SSC), 13, 14, 276
Supercooling, 267
Supergravity, 227–236, 244, 252
Superparticle, 230
Superpartner, 233
Super proton synchrotron (SPS), 196
Superspace, 230

Superstring theory, 247–260
Supersymmetry theory, 13–14, 229, 232
Supertheory, 14
Susskind, Leonard, 218, 249
Symanzik, Kurt, 155
Symmetry, 117–121
Symmetry breaking, 141
Synchrotron, 56

Tau, 3, 42, 186–187
Tau–theta puzzle, 131
Technicolor theory, 218–219
Tensor field, 134
Tevatron, 12
't Hooft, Gerard, 145–147, 223, 264
Theory of Everything (TOE), 251
Thomson, J. J., 17, 20, 24–25
Ting, Samuel, 169–171, 179
Tomanaga, Shin'ichiro, 93
Transformation theory, 35
Triplet, 66
True vacuum, 267
Tryon, Edward, 273
Tye, Henry, 267

U particle, 43
UA1 detector, 201–203
UA2 detector, 201–203
Uncertainty principle, 33, 36–37
Unification, 4, 205–216
Upsilon, 187–189

V particle, 47
Vacuum polarization, 80
Van de Graaf, Robert, 49
Van Nieuwenhuizen, Peter, 231
Vector field, 133–134
Veltman, Martinus, 145–146
Veneziano, Gabriele, 249
Virtual photon, 36, 80
Virtual pion, 94

W particle, 6, 65, 135, 137, 196–203
Walton, Ernest, 5, 49
Wave packet, 32
Weak nuclear force, 4
Weinberg, Steven, 143–144, 210, 218
Weinberg–Salam theory, 143–147, 165
Weisskopf, Victor, 81, 83–84, 98, 105
Wess, Julius, 229–231
Weyl, Hermann, 120–121
Wilczek, Frank, 155–156
Wilson, Ken, 153, 163
Witten, Edward, 255, 258, 271
World sheet, 256
Wu, Chien-Shiung, 132

X particle, 208–214, 266

Yang, Chen Ning, 68, 122–124, 131
Yang–Mills theory, 69, 124–125
Yau, Shing-Tung, 258
Yukawa, Hideki, 42–47
Yusif-Zambal, Farhid, 271

Z particle, 6, 137, 144, 202–203
Zumino, Bruno, 229–231
Zumino–Deser paper, 231
Zweig, George, 101–102

QC 173.7 .P364 1987

Parker, Barry R.

Search for a supertheory

Discarded
Date

JUN 23
2025